High-Technology Policies

High-Technology Policies

High-Technology Policies

A Five-Nation Comparison

Richard R. Nelson

American Enterprise Institute for Public Policy Research
Washington and London

Richard R. Nelson is Elizabeth S. and A. Varick Stout Professor of Social Science at Yale University. He has written numerous books and articles analyzing R&D policy in the United States and in other countries.

Library of Congress Cataloging in Publication Data

Nelson, Richard R.
High-technology policies.

1. Electronic industries—Government policy. 2. Semiconductor industry—Government policy. 3. Computer industry—Government policy. 4. Aircraft industry—Government policy. 5. Nuclear industry—Government policy. 6. High technology industries—Government policy. I. Title.
HD9696.A2N45 1984 338.4'76213817 84-21640
ISBN 0-8447-3566-3
ISBN 0-8447-3565-5 (pbk.)

1 3 5 7 9 10 8 6 4 2

AEI Studies 412

Printed in the United States of America

Contents

Foreword

High-technology industries provide the cutting edge for U.S. international competitiveness. They contribute significantly to economic growth, trade performance, and national security and play a central role in shaping the direction of progress in the national economy.

The importance of high-tech industries to the health and well-being of the U.S economy has prompted a number of industrial policy proponents to advocate government policies designed specifically to facilitate technological development and competitiveness in these industries. While it is generally recognized that competition in free markets is chiefly responsible for many of the remarkable technological advances over the past century, industrial policy advocates argue that the demands of the increasingly competitive international market will require more active U.S. government involvement in the future.

Government institutions and policies certainly have not been neutral in the technological innovation process. Many government programs and policies in the United States and other industrialized nations have been erected or advocated in an effort to accelerate the rate of technological progress. The effectiveness of these programs, however, is open to question.

Dr. Richard Nelson's monograph *High-Technology Policies: A Five-Nation Comparison* provides an important contribution to the debate over the proper role of the public and private sectors in the innovation process. The author examines technological developments in three important high-tech industries—semiconductors and computers, civil aircraft, and nuclear power—and focuses specifically on the government policies of five major industrial countries—the United States, West Germany, France, Japan, and Great Britain—that bear on these industries.

Three large questions give direction to his inquiry. Do strong high-technology industries give special economic advantages to countries? What is the principal direction of causation between changes in the economy at large and in the high-tech sector? And what types of government programs and policies have been successful, and what types have been failures and why? Dr. Nelson's analysis and conclusions illuminate many of the complexities underlying the often politically charged debate over the viability and desirability of a national industrial policy.

Dr. Nelson does not underestimate the important role MITI and similar government agencies have played in achieving technological advance for Japan and other nations. He concludes, however, that "the flexible industrial structure of the United States should not be discounted as a formidable competitive engine for progress. We may be lucky that it so stubbornly resists being targeted, coordinated, or planned."

High-Technology Policies: A Five-Nation Comparison is one of many studies sponsored by the American Enterprise Institute's multiyear project on international trade, technology policy, economic adjustment, and human capital development. Entitled Competing in a Changing World Economy, the project is designed to examine basic structural changes in the world economy and to explore strategies for dealing with the new economic, political, and strategic realities facing the United States.

WILLIAM J. BAROODY, JR.
President
American Enterprise Institute

Preface

This study considers a small but important part of the ongoing debate about industrial policies. It is concerned with the roles government can play effectively in furthering high-technology industries and the kinds of policies that are ineffective or worse. Over the past quarter century there has been a considerable amount of experience bearing on this issue, and my aim here is to recount some of that history and attempt to draw lessons. The focus is on three major industry complexes: semiconductors and computers, civil aircraft, and nuclear power. I examine the programs and policies bearing on these industries in five major capitalist nations: the United States, France, Great Britain, West Germany, and Japan.

The present study is the latest product of my long-standing interest in government policies supporting technological advance in industry. My first major venture in this arena—a book written jointly with M. J. Peck and E. D. Kalachek—focused on U.S. policies; it was published in 1967, more than twenty-five years ago. Several years later I collaborated with George Eads in a study of the U.S. supersonic transport and breeder reactor programs (1971). More recently I have worked with a group of colleagues on the study of U.S. policies in support of technological advance in seven different industries (Nelson, 1982). The present study draws heavily on this earlier work, but here I broaden the range of countries involved and narrow the range of industries.

Over the course of my research and writing on U.S. policies, I have developed a set of perceptions and arguments that significantly influenced the present study. Indeed, one reason I undertook it was to see if the generalizations held up for other countries or whether they needed significant modification in other contexts. I set these out tersely as follows.

Technological advance in industry must be understood as an integral aspect of industrial competition. It is motivated and shaped by the carrot of perspective advantage and the stick of fear of losing the competitive race. The incentives for innovation provided by this context have been uneven but powerful. They have drawn forth significant innovation in some industries, but not in others; and they have strongly motivated the attention of profit-oriented companies to certain kinds of R&D, but not to other kinds. While the competitive process has ground unevenly and obviously is rather wasteful in several respects, it has yielded extremely high social returns. Capitalism has been, and continues to be, a remarkable, if imperfect engine of progress.

While the workings of competitive private enterprise are a central part of the system of technological progress in advanced capitalist countries like the United States, governmental institutions and policies, as well as private not-for-profit ones, also have played an important role. Like the private enterprise part of the system, the governmental part is variegated. Government has played various roles, with different purposes, and the magnitude and structure of the governmental role have varied significantly from industry to industry. Given the objectives, some policies and programs have been very effective. There have been, however, several fiascoes.

The three industry complexes considered in this study all have been marked by rapid technological progress, significant private R&D efforts, and large government programs aimed at advancing technologies. The objectives of the programs and the structure of the policies, however, have differed in interesting ways from industry to industry and across the group of countries considered here.

This record is interesting to consider in its own right and for the general lessons it affords about government policies. It is especially interesting in light of the present debate about industrial policy. Several proposals have been made to the effect that the U.S. government should significantly step up its support of our high-technology industries. This essay presents and analyzes the record of past policies aimed at an important subset of these industries. It proposes to illuminate the kinds of objectives and the kinds of policies pursued that might make sense and the kinds that are likely to turn out misguided or inappropriate.

Acknowledgments

The research that led to this manuscript was supported by the Division of Policy Research and Analysis of the National Science Foundation under grant PRA-8019779. My knowledge of the industry-specific country programs in other countries was greatly enriched by research done by Yale graduate students. I am indebted to the doctoral thesis work of Franco Malerba and Donna Doane for much of what I know, respectively, about European and Japanese electronics; to William Spitz for his research on the policies of the European governments toward their aircraft industries, which was commissioned for this study; and to Michael Sullivan for his work for this study on the European nuclear programs. In addition to the many published works from which I have drawn, I have had the advantage of studying the draft manuscripts of my colleague Merton J. Peck on the Japanese programs in support of semiconductor and computer technologies. George Eads helped me to understand the vague issues of antitrust. I am indebted to Bo Carlsson, George Eads, Therese Flaherty, Otto Keck, Richard Levin, Franco Malerba, Sharon Oster, Keith Pavitt, Merton J. Peck, Robert Reich, and William Walker for reading earlier drafts and making helpful suggestions. The remaining errors are all my fault.

1
Introduction

An important part of the current discussion about industrial policies is concerned with "high-technology" industries, defined as those characterized by large research and development (R&D) expenditures and rapid technological progress. It has been proposed that, in high-income countries at least, technological change in these industries drives economic growth more generally and that government policies should explicitly aim to facilitate the progress and competitiveness of these industries. The stakes are seen as largely economic, although there may be important political values as well.[1]

Many arguments, some of them complex and subtle, have been put forth in support of this position. Not much injustice is done, however, if I paraphrase them in terms of two related, but distinguishable, propositions. The first is that high-technology industries often are "leading," in that they tend to drive and mold economic progress across a broad front. The second is that high-technology or leading industries are "strategic," in that national economic progress and competitiveness are dependent upon national strength in these industries, and governmental help is warranted to ensure this strength.

The leading industry idea has a long tradition among scholars as well as among sophisticated lay observers. The sharpest articulation probably is Joseph Schumpeter's. In his *Business Cycles* (1939) he observed that economic progress is not steady but occurs in "long waves"—an idea put forth earlier by Nikolai Kondratieff—and proposed that these were caused by periodic surges of technological innovation. He associated each upswing of the Kondratieff cycle with a cluster of innovations in particular leading industries—textiles and machinery in the first part of the nineteenth century, iron and steel and railroads in the second part, automobiles and chemicals and electrical equipment in the beginning of the twentieth century. He argued that technological advances in these industries had wide effects and, indeed, more or less determined the general economic climate of an era. The notion that the second half of the twentieth century is being shaped by innovations in electronics, particularly as applied to computation and communication, and to a lesser extent by vastly speeded long distance transport, clearly is in the spirit of Schumpeter's theory.

The leading industry notion involves some combination of significant ongoing technological advance and widespread economic effects. An industry can be leading without being particularly high-technology and clearly not all high-technology industries are leading. The three industry groups that will be the particular focus of this essay, however—electronics, aircraft, and nuclear power—all have been highly R&D intensive and have had, or were expected to have, major shaping effects on a wide range of economic activities.

The idea that high-technology, and particularly leading, industries are "strategic," in the sense that they warrant special favor and support, also seems to have been around for some time. Countries trying to modernize and catch up with a perceived leader—as Germany tried to do vis-à-vis Britain in the midnineteenth century—often have given special treatment to certain high-technology industries of the day—then steel and machine making—seeing these as source and symbol of the leader's strength. The last decade or two has been marked by increasingly sharp articulation of the idea and by the adoption in many countries of extensive policies explicitly based on it.

Although related, the hypothesis that certain high-technology industries are leading and the hypothesis that certain (generally the same ones) are strategic are not the same. The first is about the way the effects of certain kinds of technological advance fan out to influence economic activity widely. The hypothesis is silent about either the role of national boundaries or about the returns to a nation of providing special nurture to those industries. The second is exactly about these matters.

Nor does the second hypothesis follow logically from the first. One can accept the leading industry hypothesis and at the same time be skeptical about whether any major strategic advantages accrue to the countries where these industries are largely based or whether any special assistance to these industries is warranted. If international economics were as depicted in standard neoclassical trade theory, it is hard to see any general national advantage stemming from a strong position in high-technology or leading industries. Rather, that theory would reverse the discussion. The orienting question would be what kinds of factor endowments and other conditions give a country a comparative advantage in high-technology industries. The presumption would be that, given those conditions, it is advantageous to exploit that comparative advantage; otherwise it is not advantageous.

To the extent that a comparative advantage can be built through various forms of investment, the wisdom of such investment should be assessed in terms of the standard rate of return. According to this point of view, there certainly is nothing special about high-technology industries. Market mechanisms work as well, or poorly, on them as on other industries.

One can piece together two plausible counterarguments as to the reasons high-technology or leading industries are also strategic ones for high-income countries.

One argument is based on a product-cycle theory of trade, amended by a proposition that the returns to R&D are not fully appropriable by the industry that undertakes it. A starting premise is that high-wage countries need to be competitive in high-technology products if they are to be competitive in anything. Given a considerable degree of international capital mobility, high wages can be maintained and increased, only if a country has a special capability for producing things that low-wage countries cannot. In some cases these capabilities may be related to special access to certain raw materials, or to climatic advantages. Mostly, however, if high-wage countries are to be able to compete, they must be ahead of other countries in the creation and implementation of new technologies. Add to the product-cycle theory an argument that, while many of the relevant investments in new technologies are appropriately made by private individuals and business firms, some of the most important investments, specifically R&D, yield significant externalities. Then one has a case for public support of these latter investments, which indeed are of strategic importance to high-wage countries.

The argument above is not tied to the notion that high-technology industries are necessarily leading. A second and different argument is concerned explicitly with leading industries. The core idea is that, since technological advance in leading industries yields opportunities for innovation in the industries that buy from them, firms in these connected industries can reap externalities if they can exploit the new opportunities before their competitors do. The key is good information connections. If communication proceeds more effectively within national boundaries than across them, then a nation's high-technology industries indeed may lend strategic advantage to the nation's downstream industries. It may make sense, then, to subsidize or protect national firms in the key industries to obtain these interindustry externalities. Of course this argument, while different, is not incompatible with the product-cycle, general R&D externalities argument.

While these arguments are plausible, they have not been well documented empirically. At the least their quantitative importance is unclear. The question of whether high-technology or leading industries are strategic should be regarded as open.

But even assuming that these industries are strategic and that there are strong arguments for a national effort to encourage and support them, the policy implications are not immediately obvious. Most of the current policy discussion is focused on policies explicitly targeted to aid them. It can and has been argued, however, that the key to strength in high-technology industry resides in more broadly based factors. Thus, David Landis (1970), in discussing why Britain led the continent in the industries that sparked the early industrial revolution, stresses the general flexibility of British economic institutions compared with those on the continent at that time. His analysis of the reasons Germany overtook Britain in steel, chemicals, and elsewhere is posed in terms of the education and banking systems that arose in Germany but not in Britain. The American supremacy in average worker productivity in manufacturing in general and in per capita income preceded the rise of our semiconductor and computer industries. Several scholars have noted that the American system of higher education had unusual strength in the period after World War II. Many have noted that in Japan the extremely high investment rates and the development of an educational system that outstripped the United States in the production of engineers came before Japan gained strength in electronics. Keith Pavitt (1976) has argued that the ability to exploit the technological opportunities afforded by leading industries requires strong technological capabilities in a wide range of industries: chemicals, machine tools, and other metal products are good examples. If strength in the high-technology industries and their downstream partners is basically a concomitant of general and broadly based economic and technological strength, rather than a basic cause, then it may make little sense to try to stimulate these industries specifically.

There is also the related question of what, if anything, narrowly aimed government policies can do to help domestic companies get and stay in the forefront of industries in which technology is advancing rapidly and considerable international competition exists. In the view of several American writers on the subject, a policy in support of high-technology industries is to be distinguished sharply from a policy of supporting more traditional industries because they are in trouble. Indeed the proposal is to shift the emphasis of our industrial policy from the traditional to the high-technology industries and away from protection and toward more positive support. In the United States, however, the basic high-technology industries are, by now, quite traditional. Representatives of those industries, as of older ones, rail about unfair foreign competition and call for offsets of various sorts, if not for blatant protection. Proposals for positive support are less well articulated.

It seems recognized that staying in the forefront of a rapidly advancing field is not the same as closing a gap with the industrial leader. The policies of foreign governments, particularly Japan, often held forth as possible models for the United States, however, have usually been of the "catch up" sort, and may not be well suited to preserving or enhancing technological leadership.

To recapitulate, there are several basic questions about policies in support of high-technology or leading industries in the expectation of significant national economic advantage. One question relates to the gains a country reaps from being strong in the leading industries of the day. Do special economic advantages accrue to a country because it is strong in high-technology industries; and if so, what are those advantages? The second question relates to the direction of causation. To what extent does strength in these industries flow from general economic strength rather than the other way around? A third question is about the efficacy of more narrowly focused instruments. What kinds of industry-specific policies are feasible and effective, and what kinds are infeasible, ineffective, or worse.

To begin to explore these questions, in chapter 2, I review some of the salient features of the process of technological advance and of industries in which technological advance is rapid. Such a review helps to identify the opportunities for and the constraints on public policies meant to allocate resources more effectively to further technological advance. Then, in chapters 3 through 5, I turn to the actual experiences the major economic powers have had with industrial policies. This recounting is partly description but also partly analysis, since the choice of what policies to describe and how to describe them innately involves judgments about what is important. The countries studied are the United States, Britain, France, West Germany, and Japan. As mentioned earlier, three industry groups will receive special attention: semiconductors and computers, civil aviation, and nuclear power. In chapter 6, I return explicitly to the three basic questions raised above and try to provide tentative answers to them. In the concluding chapter I make some remarks about plausible directions for U.S. policy.

2
Characteristics of Technological Advance in Leading Industries

This chapter highlights certain features of technological advance in leading industries that need to be understood when one thinks about government policies to help these industries. The account draws on various studies of technological advance in aviation, nuclear power, computers, semiconductors, and several other industries. Although there are important differences in patterns of technological advance in these industries, there also are certain fundamental similarities.

These similarities are, first, that the precise path taken by technological advance is virtually impossible to predict, and there often are major surprises. Any investment in anticipation of a major breakthrough is a gamble. Second, individual technological advances seldom stand alone. Almost always they connect intellectually and economically both to earlier advances along the same lines and to advances in other but related technologies. Third, a competitive market provides a rather special structure to information relevant to R&D decision making at any time and establishes a particular set of incentives and constraints. While a competitive market may stimulate progress, it also causes certain inefficiencies and wastes beyond those inherent in the process of technological advance itself. These "market failures" are appropriate targets for public policy. Fourth, although there surely are targets of opportunity in the sense of rather obvious shortcomings of market institutions, there are limits on the ability of public policies to hit these targets. Certain constraints are caused by the implicit guide rules of market competition that limit what government actors can do in the game. Others have to do with more general limitations on the policy tools that governments can fashion to spur and guide technological progress.

Uncertainty

It is important to recognize the essential uncertainties in an industry in which technology is advancing rapidly that surround the question, Where should R&D resources be allocated? There generally are several ways in which the existing technology can be improved and at least several different paths toward achieving any of these improvements. Before the fact it is uncertain which objective is most worthwhile pursuing and which approach will prove most successful. Aviation experts, for example, disagreed on the relative promise of the turboprop and turbojet engines; those who believed in the long-run promise of commercial aircraft designed around turbojet engines differed about when to go forward with a commercial vehicle. Computer designers disagreed about when computers should be transistorized; later the subject of the extent and timing of adopting integrated circuit technology in computers divided the industry.

In a certain sense, technological advance is a wasteful process. Inevitably there is a litter of abandoned ideas and projects, some of which cost plenty. Hindsight suggests that there ought to be ways to tidy up the process, to avoid marching down false paths, to figure out in advance which technology will be best. But hindsight is better than foresight. Although some of the failed efforts strike the contemporary reader as obvious blunders, that they were so was not obvious to the people who made the key decisions at the time.

There are market as well as technological uncertainties. To judge how much merit customers will see in a radically new design is no easy task. The customers themselves may not know before they have tried out the design. The favorable public response to the smoothness of jet passenger flight was easy to underestimate, and the lack of willingness to pay for supersonic flight easy to have miscalculated. Before such machines were made available, there was no apparent business demand for computers. The value of an innovation may depend on unpredictable events, such as whether a complementary product is available, or on how the market develops for a product for which it is a component part. The post-1973 hikes in fuel costs surely hurt the supersonic transport and helped Airbus.

If the problem were simply uncertainty, but everybody agreed on the structure of the uncertainty, one could define the R&D allocation problem as being something like a dynamic programming problem involving uncertainty and learning. An optimum strategy in such a dynamic programming problem well might involve exploring various possibilities and holding off commitment to a single one until lots of evidence were acquired. I say "something like" a dynamic programming problem because in that formalism all the possible branches in the tree are assumed to be known in advance; it is their realization that is uncertain. In contrast, a well-known characteristic of R&D is that surprises occur; things happen that no one thought of and that call for a rethinking of the whole program. But if everyone saw the problem and the uncertainty in the same way, one still could think of trying to plan R&D broadly, in the spirit of dynamic programming, around the consensus of knowledgeable people. This thought has merit.

Yet a key characteristic of the R&D environment is differences of opinion and vision. Human beings, and organizations, seem to be innately limited in the range of things they can hold in mind at any time and even in the way they look at problems. Some individuals simply see things about a problem, or about an alternative, that others do not see; what is seen may or may not actually be there. But that different people look at a problem in different ways and see different things about it means that terms like insight, creativity, and genius often are applied to successful inventors or laboratories. It usually is not clear in advance to anybody in a position to make judgments about the matter just who is going to bet right this time. Committees of experts are unreliable judges of these issues, even if, or particularly if, they are forced to arrive at agreement.

The implications are important. The uncertainty that characterizes technological advance in high-technology industries warns against premature unhedged commitments to particular expensive projects, at least when keeping options open is possible. The divergences of opinion suggest that a degree of pluralism, of competition among those who place their bets on different ideas, is an important, if wasteful, aspect of technological advance.

Connectedness

Particular technological advances seldom stand alone. They usually are connected both to prior developments in the same technology and to complementary or facilitating advances in related technologies.

Many technologies advance over time in what might be called an evolutionary manner, with today's round of R&D activities aimed at improving today's prevailing technologies in certain particular directions or at creating variants better designed for particular purposes. Thus one can see in the most recent designs of commercial jet aircraft ancestral connections to the first round of commercial jet airliners—the Boeing 707 and the Douglas DC-8—created more than twenty-five years ago. While, measured in terms of the rate of performance enhancement or the reduction in cost per operation, technological advance in semiconductor memory devices has been spectacular, one can recognize a natural sequence in the generations of memory devices, from the first integrated circuits more than twenty years ago.

Evolutionary change is punctuated by revolutionary change. In civil aircraft the advent of the successful commercial jet airliner in effect changed the basic nature of airliner technology from the earlier piston engine-based regime. The integrated circuit represented a sharp break from the earlier discrete transistor, which in turn had involved a revolutionary shift in electronic device technology from vacuum tubes. These sharp shifts in technological regimes often were marked by changes in the nature of the predominant companies. Thus, as jets replaced piston-driven planes, Boeing replaced Douglas as the leader in the design and production of airliners. With the advent of the integrated circuit, the old electronic equipment producers, like General Electric and Westinghouse, failed to stay competitive and were replaced as technological leaders by such companies as Texas Instruments, Mostek, and Intel.

Technological advances often are linked together because certain products form relatively tightly integrated systems. The development of more efficient and powerful bypass jet engines in the 1960s made possible the wide-bodied jet passenger airliners. Integrated circuits are the heart of the modern computer. In a systems technology, an advance in one part of the system may not only permit but require changes in other parts. Thus a computer designed around integrated circuits is a very different machine from one designed around vacuum tubes.

The term *system* connotes a recognized strong interdependence between components. Institutionally this recognized interdependence leads either to the development of companies that design several of the key components themselves or to strong interactions, sometimes contractual, among companies producing different components.

The tightness of interdependence, and of organizational connectedness, of course is a matter of degree. When ties are relatively loose, the concept of systems connectedness becomes somewhat awkward, but some of the same phenomena show up, in weaker form, in terms of upstream-downstream connectedness. The modern jet engine would not have been possible without prior advances in metallurgy. Further progress in integrated circuits will depend on developments in the instruments that trace out the circuits.

There are several important implications of this connectedness. First, experience in a technology counts. In many modern technologies a firm must gain mastery over older or simpler aspects before it can gain competence to work at the leading edges. And firms that introduce a new product first gain learning curve advantages over their competitors, provided someone else does not come out with a significantly better design. Thus there is room for "infant industry" arguments. But it is by no means inevitable that a protected infant will grow up to be competitive.

Also, experience and competence in a particular technological regime may count for little, or be disadvantageous, when a significant shift in technological regimes occurs. A regime shift signals opportunities for new companies and requires significant changes in perceptions and policies of established companies if they are to remain competitive. This need for change may pose severe problems for an industrial policy that is committed to the support of a particular set of companies.

Second, to be successful in a high-technology industry, a firm needs to be "plugged in" to a wide range of technologies. Recognition of these interdependencies is at the heart of some arguments in favor of active national industrial policies to spur leading industries. It is an open question, however, whether the communication networks are or can be truncated at national borders. The presence of multinational corporations in high-technology industries further complicates this question. I shall return to it later.

The Competitive Market Context

Joseph Schumpeter, more than anyone else, has shaped the way scholars view competition in technologically progressive industries. Schumpeter's core message was that the most socially valuable form of competition, in capitalist economies, was through technological innovation.

Proprietary technological knowledge drives the capitalist engine. The principal ways to achieve proprietary benefit are secrecy, patent protection, and a head start. There are significant differences among technologically progressive industries in the extent to which these different mechanisms are effective. In the pharmaceutical industry, where it is easy technologically for one company to copy another's drugs, patents play an important role both as a spur for product innovation and as a protector of a company's successful products. In semiconductors, however, patents do not appear to play such an important role, in part because they are difficult to enforce and in part because a simple head start down the learning curve often gives a company a durable and profitable advantage. In industries like those that design and produce large commercial aircraft or mainframe computers the technologies are sufficiently complicated that they simply are difficult to imitate, even when they are not well protected by patents. But although the mechanisms differ, in each of these technologically progressive industries, where privately funded research and development has been substantial, through one mechanism or another firms are able to profit from their R&D successes.

The Schumpeterian system has been an extraordinarily effective engine of progress. It has shown sensitivity to changing patterns of demand by consumers. The payoff to a firm lies in producing not simply a technologically advanced product but a product that consumers will buy in quantities at a profitable price. Profitable companies and technologically progressive industries are characterized by strong market research as well as by strong R&D. At the same time competition among firms, accompanied by secrecy about just where each is laying its technological bets, willy-nilly generates a reasonable diversity of approaches to problems and of new products.

A careful scrutiny, however, either of the models that capture, in abstract form, the nature of Schumpeterian competition or of the empirical history of technological advance in any field, indicates that the portfolio generated by market competition can in no way be considered optimum. There is virtually certain to be a clustering of effort, verging on duplication, on alternatives widely regarded as promising and often a neglect of long shots that, from society's point of view, ought to be explored as a hedge. On one hand, that one company has a patent on a product or process may induce competitors to try to invent around it—an effort that may yield something really new but that often is simply wasteful duplication. The premium placed on achieving an invention first, to get a patent or at least a head start, may lead to undue haste and waste. That three companies—McDonnell Douglas, Lockheed, and the Airbus consortium—all tried to compete in the market for wide-bodied, medium-sized airliners surely meant that total costs were excessive, if it also meant that the airlines got a good deal.

On the other hand, that certain kinds of technological advances are not well protected by patents and are readily copied deters companies from investing in these, even though a significant advance would lead to enhanced efficiency or performance. Before the advent of hybrid corn seeds, which cannot be reproduced by farmers, seed companies had little incentive to research and develop new seeds, since the farmers themselves, after buying a batch, simply could reproduce them. The farmers had little incentive to do such work since each had small holdings and had limited opportunities to gain by having a better crop than a neighbor. Within an industry, problems vary in the extent to which the problem solver gains a special advantage. In an industry in which scientists and engineers are mobile it is hard to keep secret information about the broad operating characteristics of a particular generic design or about the properties of certain materials. Such knowledge is not patentable, and, even if patentable, it would be very hard to police.

Constraints and Bases for Public Policies

It is tempting to regard these kinds of "market failures" as both justifying and guiding governmental actions to complement, substitute for, or guide private initiatives. At the least the recognition of them guards against the simplistic position that the R&D allocation naturally induced by market forces is in any sense "optimal." Propositions about where and how market forces work poorly, however, cannot alone carry the policy discussion very far. In the first place, market institutions themselves constrain public policies. It is politically difficult and likely futile to try to force a policy on an industry. Second, the market-failure language ignores the fact that, in all of the major countries studied, there long has been a strong public as well as private presence in high-technology industries. These traditional policies at once represent responses to pressures to do what the market does not do and reflect a nation's broad political attitudes regarding appropriate fields of public action. They also often constitute the reservoir of experience and the acquired customs of policy that inevitably shape new departures. I will consider these matters in turn.

As noted earlier, although it is occasionally a liability, in most instances detailed knowledge of prevailing technology, its strengths and weaknesses, is a prerequisite for knowing where and how the technology ought to be improved. The business firms that actually employ a productive process to make a particular product, and their customers, thus have a major advantage in relation to outsiders in being able to see what kinds of R&D projects make sense and what kinds are likely to be worthless. In some instances interested business firms may be willing, even eager, to let others know what they think ought to be done, particularly if the firms believe they will gain no particular advantage from doing that work themselves.

But if there is a potential profit in getting ahead of competitors, a firm is unlikely willingly to make public or disclose to a government agency the way it thinks the technological bets ought to be laid. As a result, a government agency may be cut off from the most knowledgeable expertise on the question. In particular, market information may be very difficult for a government agency to obtain unless the companies want to give it. Relatedly, a government agency may be sorely limited in its ability to find out where firms are allocating their own R&D efforts. To the extent that public money aims to fill gaps in the private portfolio, finding where these gaps are may be no easy matter. There is also the danger that public funds may duplicate or replace private funds.

Also, private firms are likely to resist governmental programs that they see as cutting into their own turf or helping competitors. In a democracy industrial policies must be regarded as "fair." Put more generally, to think that an industrial policy can successfully be imposed upon an industry is a mistake. To be effective a policy requires a degree of cooperation and participation from the industry, and members of the industry inevitably are going to be influential in shaping any policy.

New policies in support of high-technology industries in search of economic advantage also are constrained and molded by the fact that they are not planted in new ground. The Schumpeterian view of technological progress and competition, sketched above, is one-sided. It highlights proprietary technologies, private institutions, and the profit and power motives of private parties and leaves hidden in the shade the considerable long-standing public involvement in high-technology industries. In many countries this involvement has intensified significantly in recent years as the new industrial policies have been consciously set in train. The modern policies, however, have recognizable roots in more traditional policies, a grounding that at once gives them a certain legitimacy and a set of habits of thought and action that may or may not be appropriate to the new purposes. It seems useful to distinguish among three admittedly overlapping areas of traditional public involvement: support of scientific and technical education and research, public (largely military) procurement, and general modernization policies. While the details and vigor of these three broad policies have differed from country to country in ways that will be described in the following sections, there are certain common elements that I will sketch here.

In the United States state governments, with assistance from the federal government, began to take major responsibility for training in the agricultural and mechanical arts as early as the midnineteenth century. Support of research in the agricultural sciences came soon after. After World War II, the federal government gradually took on primary responsibility for support of scientific and technical education and university research generally. In Germany and France there also is a long tradition of major government support for these activities. Support by the Japanese government dates from the late nineteenth century. In Britain acceptance of a major governmental role came later but was in place after World War II. The ideological bases for such support have been varied. In popular democracies like the United States, there has been long-standing acceptance of a public responsibility for broad-gauged education and training of the citizenry. In France such policies have been associated with training and support of an elite civil service. Germany since the early nineteenth century and Japan since the late nineteenth century have explicitly pushed education and science to catch up with those countries they perceived as the technological leaders.

While much of governmental support of academic research and teaching goes to the traditional basic sciences like physics, a good portion goes to the applied sciences—like pharmacology, computer science, or electrical engineering—which are quite close to certain technologies and industries. Public support partly reflects and partly ensures that technological knowledge has an important public component as well as a private one. The public part of technological knowledge generally does not relate to the design or operational details of a particular product or process, but to generic knowledge—broad design concepts, general working characteristics of processes, properties of materials, testing techniques, and so forth. Such knowledge often is not patentable. Although such knowledge sometimes can be protected by industrial secrecy, maintaining secrecy may be difficult. Also, this kind of knowledge must be imparted to those trained to be engineers or advanced technicians. Therefore, if the relevant knowledge were proprietary this would seriously interfere with the ability of technical schools and universities to provide good training. Thus strong incentives exist for treating such knowledge as public. In many fields there is a well-established research community, with participants both in universities and in industry who contribute to generic knowledge.

The presence of well-established networks of generic research, rooted in academic institutions and traditionally financed in good part by government, provides one important road into industrial policy. So long as the R&D support program sticks close to generic work, the problem of proprietary rights is partially averted. A consultative structure already stands for mapping out sensible allocations. As I shall show, however, although the traditions of such policies point to support of academic institutions, a characteristic of the new policies in support of high-technology industries is that much of the work is done by industry, not in universities or governmental laboratories.

Public procurement demands are another traditional source of public involvement in high-technology industries. For centuries sovereigns have maintained arsenals and other workshops producing the goods they needed and have concerned themselves with the adequacy of supplies of military and other items. Since World War II, in the United States, Britain, France, Sweden, and several other countries, the armed services have been major supporters of R&D in the industries from which they procure equipment. Although defense is the largest procurement interest, in several countries space agencies, telecommunications networks, electric utilities, and television networks are operated and controlled by the government and also are important sources of demand for high-technology industries.

Procurement demands, particularly if they involve national security, help break political constraints on government action vis-à-vis industry. And such public programs often are associated with direct funding of R&D in industry.

Policies in support of high-technology industries seeking economic benefits have grown at least as much out of traditional defense procurement policies as from traditional policies in support of generic research. But the technology relevant to products that a government agency wants to procure may or may not be a basis for products that will sell profitably in a civilian market. One key question is whether, and if so how, variants of the old procurement policies, more consciously aimed to enhance civil capabilities, can give the domestic industry an advantage in international competition. Another question is whether security and economic interests are complementary or whether they entangle each other.

Among the present-day major industrial powers, the United States and Britain are extreme in the view that government involvement in the detailed guiding of the economy is a danger to be avoided unless a clear-cut public interest, like national security, is involved. In other countries government guidance, protection, and support are seen as natural instruments to be used whenever appropriate to further the national interest. Part of the difference undoubtedly lies in the Anglo-Saxon and American legacy of defining freedom largely as freedom from government. Part of the difference arose because neither Britain nor the United States developed a tradition of strong state economic guidance, accompanied by protection and subsidy, as France did develop several centuries ago and Germany and Japan developed during the nineteenth century. In France, Germany, and Japan there is far less resistance to the idea that tutelage is appropriate for industries in the national interest and much looser criteria for so labeling an industry.

As noted above, even in the United States governments have

long been in the business of promoting, supporting, and protecting certain industries. In agriculture, a prominent example, R&D support was employed early. The defense-related industries are other examples. The French, German, and Japanese, however, have operated across a far broader front of manufacturing industry and have often been motivated by a zeal to catch up with the industrial leaders of the day—first Britain and later the United States. I suggested above that Japan's highly successful post-World War II policies should be understood in this light.

As I shall show, however, the constraints on government policy exist in these countries, as well as in Britain and the United States, if in weakened form. And the fundamental question remains: can the standard instruments of tutelage—government guidance, protection, and general (and recently R&D) subsidy—which can be well directed when the objective is to catch up with a leader, be effective in establishing and maintaining a domestic industry in the forefront of fast-moving technological progress?

Let me summarize. The new policies in support of high-technology industries with economic benefits as the target have clear antecedents in more traditional policies—support of scientific and technical education and generic technical research, procurement, and, in some of the major countries involved, government tutelage of industries deemed in the public interest. These traditional policies have, willy-nilly, served as starting places—but only as starting places—for the new industrial policies. The new policies face different objectives and constraints from the more traditional ones. Some specialized structures have been developed to deal with the new targets and problems. The issue now is the efficacy of these new structures for the new assumed tasks.

3
Policies Supporting High-Technology Industries: Quantitative Aspects

By now there is a considerable record of attempts by governments to spur their high-technology industries. To try to describe and analyze this experience, so that some lessons may be drawn, certainly seems worthwhile. But even simple description is no easy task. There is a serious problem about what to describe. How ought one go about characterizing a country's industrial policies? To what extent ought one consider a nation's military, science, and education support policies with expressly industrial policies? What about trade policies? What numbers are relevant? What kind of qualitative information? How much disaggregation is necessary?

To answer these questions, one really needs well worked out and verified theory of the determinants of performance in high-technology industries so that one can identify the kinds of policies likely to be relevant or irrelevant. In the preceding section I put forth not a sharp and well-tested theory but some apparently salient facts about the key processes and institutions involved in technological advance in leading industries and some rough inferences drawn from those facts. This provides me with a broad perspective on government policies and suggests roughly what kinds of policies are likely to evolve and, of these, which have promise of influencing technical progress effectively and which kinds of policies are likely to be ineffective or worse. But the theoretical lens is fuzzy, not sharp, and it may distort as well as clarify.

In this and the following two chapters I describe and analyze the policies of the five major industrial nations toward their high-technology industries from three different angles. I will, first, consider certain quantitative aspects of these programs, presenting data on R&D spending and reviewing some studies attempting to assess the returns to private and public R&D. In chapter 4, I will describe, qualitatively, in broad terms, the policies of these nations and how they have evolved. Then in chapter 5, I focus on three major industry groups—semiconductors and computers, civil aircraft, and nuclear power.

Each of these views reveals certain things but obscures others; together they provide a rich, but certainly still incomplete, picture of post-World War II experience. Available evidence and plausible inference do, I believe, enable one to discern at least the outlines of what the policies have been—no trivial issue in view of the several conflicting statements about them. While certain things can be confidently said about the effects of these policies, however, many puzzles, blank spots, and open questions remain.

Attempts to Measure Policies and Their Impacts

To begin, it is useful to review the data on differences and similarities across nations in patterns of total and government R&D spending. For some time the countries in the Organization for Economic Cooperation and Development (OECD) have been collecting and publishing R&D statistics that are roughly comparable. These numbers apparently enable us to assess, to a first approximation, the magnitudes of the R&D resources governments are investing in policies in support of their high-technology industries. Of course, government spending on R&D is, at best, a partial measure of government policies. Other aspects of government policies however—for example, the tax treatment of private R&D expenditures; the nature of the patent laws; the characteristics of the regulatory structures; the strength of protection; or the extent of subsidy of investment in new plant and equipment—are more difficult to measure. Measurements are likely to be less comparable across countries than the R&D data.[2]

Table 1 presents total R&D as a percentage of gross national product for our six large industrial nations, 1963–1980, and breaks down the total into defense- and nondefense-related spending. Notice the initial large U.S. lead in total R&D and the subsequent convergence of R&D intensity of the major industrial powers. Notice also that the early U.S. lead was due mainly to our large defense R&D budget; in recent years, if one excludes defense, the United States has spent less on R&D as a percentage of GNP than have Germany and Japan. An important question to explore, therefore, is how defense R&D differs from nondefense R&D.

Most of defense-related R&D is funded by government and undertaken by business firms. While space and industrial policy R&D also channels funds to industry, defense R&D generally accounts for

TABLE 1

RESEARCH AND DEVELOPMENT EXPENDITURES AS A PERCENTAGE OF GROSS NATIONAL PRODUCT, 1963–1980
(percent)

	1963	*1967*	*1971*	*1975*	*1980*
United States					
Total	2.90	2.90	2.60	2.30	2.45
Defense	1.37	1.10	0.80	0.64	0.57
Other	1.53	1.80	1.80	1.66	1.88
United Kingdom					
Total	2.30	2.30	2.10[a]	2.10	1.83
Defense	0.79	0.61	0.53	0.62	0.72
Other	1.51	1.69	1.57	1.48	1.11
France					
Total	1.60	2.20	1.90	1.80	1.83
Defense	0.43	0.55	0.33	0.35	0.41
Other	1.17	1.65	1.57	1.45	1.42
Germany					
Total	1.40	1.70	2.10	2.10	2.27
Defense	0.14	0.21	0.16	0.14	0.12
Other	1.26	1.49	1.94	1.96	2.15
Japan					
Total	1.30	1.30	1.60	1.70	2.04
Defense	0.01	0.02	—	0.01	0.01
Other	1.29	1.28	—	1.69	2.03

NOTE: Dash = data not available.
a. 1972

SOURCES: OECD, except for 1980 numbers taken from table 1 of *Technical Changes and Economic Policy*, OECD, 1980.

the lion's share of government funding of industrial research. Table 2 presents data on the share of total R&D done by industry and the share of that financed by government in the countries listed. Although the ratio of industrial to total R&D varies somewhat across the countries, the range is relatively narrow. There are differences, however, in the fraction of industrial R&D financed by government. There is strong correlation of governments' share of financing of industrial R&D, with the emphasis on defense R&D. Japan and Germany, the

TABLE 2

INDUSTRIAL R&D SPENDING, 1963–1980

(percent)

	1963	*1967*	*1971*	*1975*	*1980*
United States					
% of total R&D	68.5[a]	66.8[b]	66.8[c]	65.9	68.8
% financed by government	57.6	51.1	41.8	35.6	31.8
United Kingdom					
% of total R&D	64.5[a]	64.8	63.2[c]	62.3	66.2[d]
% financed by government	33.8[a]	29.4	33.1[c]	30.9	29.2[d]
France					
% of total R&D	48.7	51.2	56.2	59.6	59.8
% financed by government		40.3	31.5	28.0	21.6
Germany					
% of total R&D	66.0[a]	67.0	67.4	66.5	72.3
% financed by government		17.4	18.2	17.9	18.2[e]
Japan					
% of total R&D	64.6	62.5	66.5	64.3	65.3[e]
% financed by government			2.0	1.7	1.4[e]

a. 1964 c. 1972 e. 1979
b. 1968 d. 1978

SOURCE: OECD.

countries with the smallest fraction of national R&D given to defense purposes, are at the bottom of the list regarding the government's share of industrial R&D. While Germany is close to the pack, government industrial R&D spending in Japan is very low compared with the rest.

Table 3 presents data on the distribution by industry of industrial

TABLE 3

SECTORAL DIVISION OF R & D FUNDING, 1967, 1975, and 1980

(percent)

	United States			*United Kingdom*			*France*			*Germany*			*Japan*		
	Ind.	Gov.	Total	Ind.	Gov.	Total	Ind.	Gov.	Total	Ind.	Gov.	Total	Ind.	Gov.	Total
Electronics-electrical															
1967	20.0	28.8	24.4	22.3	27.9	24.1	22.7	25.6	24.6	25.2	29.8	25.9	24.4	33.0	24.5
1975	20.9	30.4	21.8	20.5	34.5	26.0	27.0	35.7	31.7	30.0	31.0	29.9	26.0	32.3	26.1
1980	19	28	22	18[b]	46[b]	26[b]	22[a]	28[a]	26	28[a]	27[a]	28[a]	25[a]	20[a]	25
Chemical															
1967	21.0	2.8	11.8	21.0	1.1	14.7	27.4	3.7	19.0	33.2	4.3	28.5	27.1	11.0	27.0
1975	21.4	3.2	14.6	29.5	1.9	19.7	26.1	2.9	19.2	35.0	2.3	29.1	22.4	2.9	22.1
1980	19	4	15	30[b]	1[b]	19[b]	26[a]	6[a]	19	27[a]	9[a]	24[a]	21[a]	5[a]	23
Machinery															
1967	17.3	6.4	11.8	14.4	7.4	11.8	7.7	2.4	5.6	12.2	37.1	16.2	10.7	22.0	10.8
1975	21.8	6.7	18.7	11.3	1.9	7.9	7.0	1.4	5.2	13.0	20.7	13.9	9.9	7.4	9.8
1980	27	7	20	16[b]	6[b]	36[b]	10[a]	3[a]	10	19[a]	14[a]	18[a]	14[a]	10[a]	14
Air and space															
1967	14.5	56.8	35.8	7.1	61.0	25.3	8.0	66.1	28.8	0.9	24.9	5.0	*	*	*
1975	8.3	54.7	24.4	5.0	58.8	23.9	6.6	57.8	20.2	2.0	40.9	9.5	*	*	*
1980	9	52	23	6[b]	46[b]	20[b]	10[a]	60[a]	19	6[a]	34[a]	6[a]	*	*	*

(*Table continues*)

TABLE 3 (continued)

	United States			United Kingdom			France			Germany			Japan		
	Ind.	Gov.	Total	Ind.	Gov.	Total	Ind.	Gov.	Total	Ind.	Gov.	Total	Ind.	Gov.	Total
Other transport															
1967	12.6	4.5	8.6	12.4	1.3	8.5	13.7	0.5	8.6	14.9	1.8	12.6	12.5	22.0	12.5
1975	13.9	4.1	10.4	12.3	2.2	8.6	15.9	0.5	11.1	14.0	0.6	11.6	18.3	50.0	18.9
1980			12	14[b]	2[b]	7[b]	18[a]	0[a]	13[a]	16[a]	4[a]	14[a]	18[a]	58[a]	18
Basic metal															
1967	4.9	0.3	2.6	7.1	0.7	5.0	6.1	1.3	4.4	9.8	0.8	8.4	10.6	6.0	10.6
1975	4.5	0.3	3.2	5.9	0.2	3.8	5.4	0.7	4.1	3.0	2.1	3.1	9.5	4.4	9.4
1980			4			3	4[a]	1[a]	4[a]			4			9
Chemical-link															
1967	5.1	0.3	2.7	9.9	0.3	6.7	10.1	0.2	6.1	2.4	0.8	2.1	7.7	0.0	7.7
1975	4.4	0.5	3.6	10.8	0.3	7.1	8.9	0.5	6.2	2.0	1.3	2.0	6.4	1.5	6.3
1980			4			6			6			3			
Other manufacturing															
1967	4.6	0.1	2.3	5.8	0.3	3.9	4.3	0.2	2.9	1.4	0.5	1.3	7.0	6.0	6.9
1975	4.8	0.1	3.3	4.7	0.2	3.0	3.2	0.5	2.3	1.0	1.1	0.9	7.5	1.5	7.4
1980			3			3	3[a]	1[a]	3[a]	2[a]	1[a]	2[a]			

NOTES: Ind. = industry; Gov. = government; * = included in "Other transport."
a. 1969
b. 1978

SOURCE: OECD except for 1980, numbers taken from Table II of *Technical Change and Economic Policy*, OECD, 1980.

R&D spending broken down by source of finance. In all countries the electronics-electrical complex of industries attracts between a fifth and a third of both private and public industrial R&D funding. Although the countries are roughly similar in the fraction of the governments' industrial R&D budget going into these industries, that in Japan and Germany the government accounts for a relatively small share of total industrial R&D means that the public share of electronics R&D financing in these countries is small compared with that in countries, like the United States, with large defense R&D efforts. Most of government R&D spending in these industries is for defense. Programs in support of reactor development channel funds into the electrical equipment industry, but these funds are relatively small. As I shall show, programs in support of commercially oriented R&D in electronics are small compared with defense-related programs.

R&D in the air and space industries is largely financed by governments. In the late 1960s such spending was closely tied to defense. The increase in Japanese and German public R&D funds going into air and space since the late 1960s is associated with some rise in their defense R&D budgets and an increase in their funding of R&D on commercial aircraft.

Considerable public R&D goes into these two large industry complexes in all countries. In Germany and Japan considerable public finance of R&D also goes into the machinery industry. The other large R&D-intensive industries, chemicals and chemical-linked products (largely pharmaceuticals) and "other transport" (largely automobiles) are financed mostly by industry.

Table 4 describes the distribution of government R&D by social objective. Notice the small percentage of government R&D going to "industrial growth, not otherwise classified." If one adds in transport and telecommunications the numbers still are small. In some countries energy-related R&D is significant; most of this R&D is nuclear power. In a few instances civil aeronautics is significant. The dominant impression, however, is of the limited scope of government R&D support for high-technology industries for commercial purposes.

But it is dangerous to draw any quick conclusions about the unimportance of government support, and there are reasons to suspect that the numbers above purporting to measure a country's active industrial policy may not tell us much. As shown in the table the two countries with the greatest commitment to government-financed defense and space R&D spending—the United States and the United Kingdom—put relatively little government money into R&D programs explicitly labeled as industrial growth. The defense and space

TABLE 4
PUBLIC R&D SPENDING, BY OBJECTIVE, 1971, 1975, and 1980
(percent)

	United States			*United Kingdom*			*France*			*Germany*			*Japan*[a]	
	1971	1975	1980	1971[b]	1975	1980	1971[b,c]	1975	1980	1971[b]	1975	1980	1975	1979
Defense	52.2	50.8	47.0	46.2	52.9	59.4	38.0	32.6	40.9	21.3	17.6	14.2	3.8	3.6
Space	19.2	14.5	14.4	1.9	2.5	2.3	7.0	6.1	5.0	9.4	6.8	6.0	11.8	9.3
Civil aeronautics	3.1	1.6	1.6	14.5	8.2	3.4	7.0	6.7	2.4	3.6	2.6	2.3	—	—
Industrial growth n.e.c.	0.6	0.4	0.4	4.6	3.1	3.4	7.0	8.9	7.6	8.6	9.1	11.7	17.7	13.9
Agriculture	1.9	2.2	2.2	2.9	4.8	4.5	4.0	4.2	4.3	3.1	3.0	2.6	22.2	18.4
Production of energy	3.6	7.1	11.8	7.5	7.1	7.3	8.0	9.4	8.5	16.4	16.8	20.1	12.8	17.8
Transport, telecommunications	1.6	1.8	1.1	0.9	0.7	0.7	6.0	3.2	3.2	0.9	2.3	2.9	3.2	2.2

Urban and rural planning	0.4	0.5	0.4	1.2	1.7	1.1	6.0	1.6	1.5	0.8	1.8	2.1	1.0	1.9
Earth and atmosphere	1.5	2.0	2.0	0.3	0.8	0.9	6.0	3.3	3.3	2.3	2.8	3.9	1.4	1.9
Health and welfare	12.2	14.8	15.2	2.8	4.1	3.9	4.0	6.5	7.6	11.6	14.5	13.9	9.7	8.3
Advancement of knowledge n.e.c.[d]	3.3	4.3	3.9	17.2	14.1	13.0	19.0	17.0	15.2	22.0	22.7	20.2	2.8 13.6[e]	2.5 20.2[e]
Total specified R&D funding	100.0	100.0	100.0	100.0	100.0	100.0	100.0	100.0	100.0	100.0	100.0	100.0	100.0	100.0

NOTE: n.e.c. = not elsewhere classified.
a. Government intramural only, except for advancement of knowledge and industrial development.
b. Not strictly comparable with following years.
c. Rough OECD estimate.
d. Excludes public general university funds throughout and for the United States and Japan also excludes basic research supported by mission-oriented agencies. "Adjusted" U.S. figure might be about 15 percent in 1980.
e. Total university receipts from government for specified projects including those for other objectives.

SOURCE: OECD, numbers taken from table 9.2 of Christopher Freeman, *The Economics of Industrial Innovation,* 2d ed. (London: Frances Pinter, 1982).

R&D funds naturally flow to the leading industries. The countries without a large defense or space program apparently have partially compensated by devising explicit R&D support programs associated with an industrial policy, at least for their electronics industries.

It seems important to know in what ways funds that are earmarked for industrial growth are allocated differently from funds that are earmarked for defense or space. Although certain gross differences seem obvious—in most cases items procured by the military differ in significant ways from items that are sold on commercial markets—there may be less difference here than meets the eye. First, what is learned in a program aimed to design and develop a piece of military equipment may lead to a follow-on product for the civilian market. As we shall see, there are a number of examples of this sort. Second, a portion of defense- and space-related R&D is not tied up in work on particular designs but is much more generically oriented. Apparently a considerable share of the R&D financed by governments in pursuit of the goal of industrial development also is generic in nature. To what extent then do defense-oriented programs and industrial-development-oriented programs finance much the same thing? It clearly is important to get behind the data and examine the programs in more detail.

The same kind of difficulties should make one skeptical about what can be learned from studies meant to measure the effect of government R&D spending. The problems are most severe when the analysis is conducted at a quite gross level and diminish somewhat when the analysis is more detailed and microscopically focused.

Cross-country analysis of the relationship between public and private R&D spending and the growth of labor productivity or total factor productivity is delicate and tricky. Simple regressions are not likely to tell us much. In the first place, the United States, until recently the clear leader in both total and public R&D as a fraction of GNP, was also by far the country with the highest labor productivity and per capita income. It is also apparent that in most industries U.S. technology was in the forefront. Thus other countries had the advantage of being able to learn from the United States. For a country trying to catch up, a little R&D may go a long way, and the level of educational attainment and the rate of physical investment may be the more important driving variables.

Thus Japan, initially the laggard of the group in terms of productivity levels, has experienced by far the most rapid growth of productivity. Until recently Japan has not spent much on R&D, but its rate of capital growth has been much faster than the other countries in the comparison group. Since the early 1960s it has stood high in the group in average years of educational attainment of its work force. Given its initial low start, however, despite its rapid growth rate by 1980 Japan still lagged behind Germany, France, and the United States in average productivity and income.

In the early 1960s Britain, France, and Germany were quite close in levels of per capita income and average productivity. Since then the capital-labor ratio in France and Germany has grown much faster than in Britain or in the United States and so has output per worker. By 1980 France and Germany, but not Britain, had come close to catching up with the United States in productivity levels.

Clearly the relationships are complicated. I believe in the analysis sketched above, but reliable quantitative estimates of the role of R&D and other factors are hard to devise.[3]

The analytic difficulties diminish somewhat, but remain severe, when the analysis is concerned with data at the industry level. Unfortunately, to my knowledge there has been no study tracing the relation between various measures of technological progress in an industry in different countries and various kinds of R&D input in those countries.[4] Virtually all studies using industry-level data have focused on the United States and have been concerned with cross-industry comparisons. They have attempted to explain the cross-industry differences in some measure of technological progress, usually growth of total factor productivity, by R&D, broken down in various ways, and by other variables. An important finding of many early such studies was that an industry's growth of total factor productivity was strongly influenced both by R&D done in the industry and by R&D done by supplying industries.[5]

These early studies usually did not distinguish between government-financed and privately financed R&D. More recent studies have. Various functional forms have been explored. In some treatments government R&D and private R&D are treated as having independent effects upon the rate of growth of productivity. In other studies government-financed R&D is treated as enhancing the effectiveness of privately financed R&D. Virtually all of the studies that treated public and private R&D as if their effects were independent found that, while the influence of private R&D on growth of total factor productivity was large and statistically significant, the estimated effect of government R&D was negligible and insignificant. The studies using a format that assumes interaction have been yielding mixed results.[6]

It obviously is important to gain an understanding of the routes through which government-financed R&D influences technological advance. The paths are unlikely to be the same in all industries, and distinguishing among different kinds of R&D support may be wise. Thus government support of R&D on agriculture is different in form and purpose from government support of R&D on a new missile. Also sensitivity to measurement problems seems important. Much of government-financed R&D goes to defense or space (or to health) and results in radically new products. It is not easy even to specify just how output should be measured in the relevant industries so that technological advances can be characterized as enhancing productivity, and it is apparent that actual productivity measures are hopelessly inadequate for getting at the effect of such technological advances. Moreover, the statistical analyses done thus far beg the question raised above—does it matter, or it does not matter, whether government R&D funds flowing to the electronics industry are part of a defense program or part of an industrial policy? Although we now understand somewhat better the nature of the government programs associated with various government R&D flows to industry and have a stronger appreciation of how the activities financed by those funds interact with other activities in influencing technological advance, we are not yet in a position to specify the form of the equation to be fitted.

There are several quantitative studies in which these problems have been avoided because the focus was on a particular narrowly defined area, on a program, or even on a project. Almost all of these studies have been of the effects of publicly supported agricultural R&D in the United States. The largest group of these has been concerned with estimating the returns of a flow of public R&D investment, often accompanied by private ones, aimed at creating a new kind of agricultural input (hybrid corn seeds) or improve a particular product (poultry). These studies have been detailed enough so that the relations used to permit estimation of a social rate of return have considerable plausibility. The estimated returns have generally been very high.[7]

There also have been some studies that have examined, in some detail, the contribution of NASA R&D to technological advance of importance to the civilian economy. Although civilian benefits usually were not the principal objective, for some of the projects studied the civilian benefits were substantial.[8]

The brief survey of quantitative research reveals both the difficulties and the promise of this line of work on the question of the efficiency of governmental R&D support. Cross-country or cross-industry studies have not yet been done with sufficient care and delicacy regarding measurement and specification to lend confidence to the quantitative results. The detailed microscopic case studies are more persuasive; but they describe only a few small pieces of the terrain, and it is hard to tell if they are representative. Increasing the number of careful quantitative case studies will both provide a better check on representativeness and help inform efforts at more macroscopic analysis regarding measurement and specification. Some of the qualitative case studies presented in chapters 4 and 5 also might help in this regard.

4

Qualitative Characterization of Broad Policy Positions

A different view of industrial policies is contained in broad qualitative characterizations of them. These characterizations are an art form that has been used to good avail by Raymond Vernon (1974) and more recently by Jack Baranson and Harold Malmgren (1981), Ira Magaziner and Robert Reich (1982), and John Zysman and Laura Tyson (1983). Sometimes, as in the forementioned studies, the analysis is explicitly comparative, country versus country. In other cases, the focus is on a single country, with other countries being treated as bench marks.

Such analyses try to identify similarities and differences and to assess the consequences of the observed differences. The latter exercise is especially difficult. And because of limitations in our ability to evaluate consequences of various differences, even the first part of the exercise—simply identifying the relevant differences—becomes problematic. The policies and institutions of the different countries that conceivably could bear on the performance of their high-technology industries are extremely rich and variegated.

There are many strategies I could follow for presenting a broad comparative analysis. For the purposes here, it seems convenient to proceed first by sketching the situation in the United States. I then describe what I think are the salient differences between the United States and the major European countries and among the European states. Finally I turn to Japan.

A Bench Mark: The American Experience

During the heyday of the fifties and early sixties, American economic predominance often was characterized in two different dimensions.[9] One was in terms of higher productivity levels, in the economy as a whole or in certain broad sectors like manufacturing or in particular industries like aircraft production. The other was in terms of more narrowly defined technological competences, as ability to produce the most advanced semiconductor or aircraft, significantly before other countries. By 1980 the U.S. lead had eroded in both dimensions. Several other countries had crept close to the United States in average worker productivity in manufacturing; and, given the vagaries of the international productivity comparisons, Switzerland, Sweden, and Germany probably should be regarded as now virtually even with the United States. The United States had lost its lead in most areas of consumer goods electronics to Japanese firms. But at the high end of the high-technology spectrum—civil aircraft, computers, and semiconductors—American companies generally continue to be world leaders. In all of these areas we remain, by far, the largest producer and the largest net exporter. This situation remains even though, in the views of Baranson and Malmgren and of Magaziner and Reich, the United States has not had a coherent policy in support of its high-technology industries whereas several of the European countries and Japan have developed such policies.

But this proposition again flags the problem of identifying what is an industrial policy and what is another kind of policy. The United States certainly has not been passive regarding its high-technology industries. In the first place, for many years the United States was far ahead of the rest of the world in terms of the fraction of its youth who went through secondary education and college. Although in recent years Japan has surged past us in engineering education, if numbers of students gaining a degree be the index, the United States continues to rank high in the fraction of the entering work force with a college degree in science or engineering. Virtually all the secondary education and the lion's share of the advanced education have taken place in public institutions and have received large influxes of public funds. Scientific and engineering education has been singled out for special help.

The United States, like other countries, came out of World War I impressed with the importance of certain high-technology industries for national security. During the interwar period various measures were taken not only to procure new military aircraft directly but to build up the technological strength of the industries producing airframes and engines. I shall give more detail on this experience later in this study. I note here, however, that the Pratt and Whitney aircraft engine company was formed with considerable governmental encouragement. There was similar encouragement, and governmental restructuring, of the radio industry. The Radio Corporation of America was formed, under governmental prodding, to increase American strength in radio technology and to cut through certain tangles about patents; the express purpose was to push our industry to the forefront of radio technology.

Since the late nineteenth century the Department of Agriculture has supported research and development relevant to farming. In the nineteenth century farming was not a high-technology industry. By World War II, however, American farming was becoming such, and the embarrassing productive success of American agriculture in the postwar era must be ascribed, in good part, to the effectiveness of what is probably the longest-lived program of government support of R&D relevant to an industry's technologies for economic purposes. Also the National Institutes of Health, which sponsor basic and generic research relevant to health and medicine, came into existence before World War II. The NIH system since World War II has provided significant support to our pharmaceutical industry through the basic research and training of scientists it has provided. Since the late 1960s, however, federal support of science and engineering education in the United States has fallen off, at the same time government support has increased in Japan and the Federal Republic of Germany.

World War II and its immediate aftermath brought several important additions to the scene. First, with the establishment of the National Science Foundation the federal government took on acknowledged responsibility for the funding of basic scientific research in the United States, at least that undertaken at universities, and for providing encouragement and support for the training of scientists and engineers. Second, although before World War II defense R&D support and other means of encouraging technological capability in relevant high-technology industries were piecemeal and sporadic, after World War II the Department of Defense systematically funded R&D in aircraft, engines, and electronic systems. Department of Defense programs were directly responsible for American preeminence in electronic computers, semiconductors, and jet passenger aircraft. Later NASA funding provided support to roughly these same industries. The role of the Atomic Energy Commission in sponsoring the development of civilian power reactors was strongly linked in the early days with its role in the development of nuclear weapons and nuclear reactors for submarines and aircraft carriers.

The presence of the U.S. policies described above and a general self-confidence of the American people that the United States was the technological and economic leader has, until recently, restrained any major moves toward the development of policies in support of high-technology industries expressly for economic purposes. There have been several episodes, however, in which such policies were seriously discussed at high levels in government. During the Kennedy administration, proposals for a civilian industrial technology program were put forth. For the most part the suggested programs were aimed not at high-technology industries but at lagging ones. In any case, not much came of this discussion. Another discussion surfaced in the first Nixon administration as a result of the fall-off in military and space R&D spending that occurred during the late 1960s and the growing apprehensions that other countries were gaining on us. Again, very little came out of this endeavor. During the late 1960s, however, the United States did mount a program of government support for the development of a supersonic transport, although the program aborted. During the Carter administration a Domestic Policy Review was organized with the purpose of identifying government policies that could help spur industrial innovation. That discussion also did not get very far, and the proposals that did emanate from it were, in effect, "zeroed" when the Reagan administration came to power.

At present the discussion is mounting again. I turn now to consider the European and the Japanese experiences with active industrial policies as they can be described at this broad level of discourse.

The European Experience

As might be expected, the European experience differs considerably from country to country.[10] I will use France as a bench mark and then discuss policies in Great Britain and West Germany.

France. French attitudes and expectations about the appropriate economic role of the government and of the relations between government and business of course differ significantly from the American. The tradition of a strong civil service actively engaged in encouraging, protecting, and subsidizing particular enterprises goes back to the Bourbons. It was not unnatural, therefore, for the French to assume that the government should play a major role in guiding industrial redevelopment after World War II.

As some of the more basic and obvious measures of reconstruction were completed, old habits of thought, newly reinforced, turned toward planning long-range economic growth. A quite detailed economic plan, drawn up in dialogue between civil servants and people from industry, became the symbol if not necessarily the substance of French industrial policy. The direction of French policies came to be fought about in connection with the formulation of the plan. The planning bureaucracy became an important voice arguing that France must modernize. Although since the late 1960s the plan and the planning bureaucracy have faded from the scene, the thrust toward modernization proved infectious and durable.

Many influential French citizens came out of the war with a strong sense of French economic as well as military inferiority and a determination to catch up. While the explicit planning structure was a new departure, the instruments of industrial policy were the traditional ones used in heightened degree. These included access to low-cost credit; outright subsidy of certain kinds of activities; protection from imports; and, in many cases, locked-in government procurement. Bank finance in France is rationed to a far greater extent, and the influence on the banks by the French government is much tighter, than in the United States. Zysman (1983) has presented a powerful argument about how the nature of a nation's investment financing system affects the ability of the government to steer allocation of funds. In France and Japan the system is amenable to effective government steering. Also, in France the government-controlled market has extended far beyond military equipment. France came out of the war with a sizable nationalized sector. In France, as in many other European countries, utilities like electricity generation and transmission, the telephone system, the railroads, and the airlines, which in the United States are private but regulated, are nationalized. This situation naturally has given to the French government a broad range of markets that could be guaranteed to French firms, although the individual public agencies might balk and claim independence. The French also have engaged in selected intervention in industrial structures. Indeed the French government has been tinkering with the structure of its electronics industry and its steel industry almost incessantly since the end of World War II.

Since the 1950s the French have been especially concerned about the adequacy of their high-technology industries. From early in the post–World War II period, French national security objectives have included not only a formidable military capability but also an ability to preserve or build that capability independently of constraints that might be laid down by Americans. These objectives led France to rebuild its aircraft design and production capabilities, develop the associated electronics, and move into nuclear weaponry, with reactor design as a byproduct. All of the standard French instruments of industrial policy—procurement, protection, and subsidized investment—and, in addition, heavy R&D support were used to build and maintain these industries.

French policy regarding its computer industry, which will be discussed in more detail in the next section, is an archtypical case. Current policies clearly show their origins in French frustration at the refusal of the United States in 1963 to sell France a large computer needed for its nuclear programs. Before the French government decided on its response, France's second largest computer company (after IBM), the formerly French-owned Machines Bull, had financial difficulties and came under the control of General Electric. When the government's plan to create a self-sufficient French computer capability was enacted, Machines Bull was excluded from the consortium of French firms put together to form a "national champion." Clearly a lot more was driving that policy than a striving for simple economic gain.

In the 1970s the notion that France's economic future rested on its high-technology industries began to take hold; it has been trumpeted by the Mitterrand government. As my earlier statistical analysis showed, the bulk of government funding of industrial R&D in France continues to be channeled through defense agencies. Over the years, however, the French government has developed various instruments to enable it to share the cost of commercial industrial R&D projects. When it came to power, the Mitterrand government had every intention of using these instruments heavily. In addition, several important high-technology firms that were still private were nationalized with the objective of gaining more government control over their R&D and investment policies.

In sum, the contemporary French policy in support of its high-technology industries for economic objectives remains intimately intertwined with its national security policies. Both long-standing beliefs that government should direct industry when the stakes are high and the national security interests in high-technology industries have led the French government to try to make detailed decisions about what fields of technology to push and even about what particular designs to develop.

Britain. Perhaps Britain can best be understood as a mixture of American and French elements. Like the American, and unlike the French, the British heritage is not congenial to government planning or directing of economic activity. Like both the United States and France, Britain came out of World War II with a commitment to maintain a strong defense establishment. And like France and unlike the United States, Britain suffered from a sense of economic inferiority, in relation first to the United States and later to several other countries. The commitment to an adequate defense capability, together with concern about economic backwardness, has led Britain into periodic flirtations with various industrial policies; but, in contrast with the French, the British always seem to have been of two minds about these policies.

The vast bulk of government industrial R&D support in Britain comes from the defense budget. In the fields of nuclear power and commercial aircraft, however, the British government has been the principal source of funding and has closely guided the evolution of technology. There have been periodic surges of subsidization of R&D on computers and semiconductors, with commercial prowess the objective. The British reactor program is generally regarded as an expensive failure. While there are a few exceptions, virtually all the airliners designed and produced in Britain and France have lost money. The Airbus, which I shall discuss later, is an exception. Britain also is similar to France in that its basic public utilities are nationalized. Thus the British airlines can be pressured to buy British-made planes and the electrical network to buy British reactors. Telecommunications can be urged to buy British-made electronic equipment. The British have not protected their electronics industry, however, nearly so insistently as the French have theirs.

While the British have been much concerned with national security and, where plausible, have preferred to make military equipment at home, they have had nothing like the French paranoia about dependence on the United States. The British generally (not always) have been willing not to develop a national capability, if it were judged very costly to do so and if a deal could be worked out with the Americans.

From time to time, generally but not always under the auspices of a Labour government, Britain has toyed with the rhetoric of general economic planning; however, the rhetoric never has amounted to much. Zeal for nationalization of key industries has waxed and waned. Nuclear power, aircraft, and to a lesser extent electronics aside, efforts at industrial reconstruction have largely been directed toward industries that were in deep financial trouble and that experienced serious unemployment with more threatened.

In Britain there has been a long tradition of broad governmental concern for the R&D activities of firms and of government encouragement and occasional support. Shortly after World War I, several English statesmen called attention to the fact that Britain had lost or was losing its technological leadership in most industries to the United States or Germany. As one way to get back into the race, a system of cooperative research associations was established with government's providing a significant share of the initial money. Britain long has had a collection of national laboratories and research centers. The National Research and Development Corporation, established in 1949, aimed to help commercialize inventions that came out of that network. A parade of ministries has been charged with beefing up the commercial technological prowess of British industry.

As part and parcel of long-standing concerns about British technological backwardness, the British educational system periodically has been discussed as a part of the problem. Compared with the American and German and now the Japanese educational systems, the British system turns out very few engineers. Several attempts at reform have each led to frustrated resignation.

In summary, the right word to describe British policies probably is schizophrenic. On the one hand, there is a long-standing bias against detailed government involvement in guiding the civilian economy. On the other hand, the British government has taken a very active and directive role in nuclear power and civil aviation and has flirted periodically with the idea in electronics.

West Germany. German post–World War II policies in support of high-technology industries differ significantly from those of France and Britain. Perhaps the major reason is that Germany does not now have and still does not aim for a major defense design and production capability. Like Britain it has not viewed dependence upon the United States for certain technologies as cause for embarrassment or alarm.

Before World War II, German governments seldom were shy about pushing an industry or an industrial development that they thought ought to be advanced for the national good. In this sense, the German tradition had been quite like the French. Government policies to support the development of industrial strength were explicitly justified by the objective of building military strength. Since World War II, the attempts of German governments to direct resource allocation have been quite constrained. The contrast with the pre–World War II view of the role of government certainly is partly due to self-conscious efforts, monitored by the victorious allies, to distance Germany from earlier traditions that had culminated in two world wars. In any case, inducements to government direction are diminished when there is no defense industry to support and no desire to build one (although Japan is a counterexample). Recently Germany has moved to engage in some military production but, for obvious reasons, this movement has been restrained.

Although postwar Germany has been touted as a bastion where market forces reign and the government does not try to plan or direct, this statement is something of an exaggeration. In the reconstruction period there was a considerable amount of government guidance and tripartite discussion about appropriate directions. Later the German government developed a strong sense for regional economic problems, and it has mounted various policies to redevelop regions appearing to be in trouble. A significant program has supported power reactors. More recently the government has consciously provided special R&D support to the computer and semiconductor industries. The Germans participate in the Airbus project. But certainly in comparison with France, Germany has done far less of picking particular industries for special government encouragement and support.

Its traditional policies of strong support of scientific and technical education and research have been sustained, however. From the days of Frederick the Great, Prussian and later German governments have strongly supported scientific and technical education. Originally the motivation was to establish a cadre of civilian and military government officials that could lead Germany out of economic and technological backwardness. By the midnineteenth century Germany was strong, even leading, in several fields of science, principally those connected with chemistry. The government actively encouraged consultation between German academic scientists and the newly founded science-based companies. In the late nineteenth and early twentieth centuries, government funds helped establish and sustain several laboratories concerned with applied R&D as well as the basic sciences. Many scholars have attributed Germany's rise as an economic and technological power during the last part of the nineteenth century to the effectiveness of those policies. By the 1920s and 1930s German industry had clearly established a position as a world technological leader in most fields of chemistry, electronics, machinery, and aviation. Its system of scientific and technical education and basic research was widely regarded as preeminent.

The traditional policies have been reaffirmed in the post–World War II era. Strength in scientific and technical training has been stressed, and the government-supported laboratory structure has been extended.

Perhaps the most interesting part of the German industrial policy apparatus is the Ministry for Research and Technology (BMFT) formed in the early 1970s. It stands separate from, not joined with, the Ministry of Economics and is focused on enhancing the technological competence of German industry. The ministry has come to act as a sort of National Science Foundation for industry. Within certain broadly defined areas, companies submit proposals to the ministry for evaluation by a committee consisting of government and nongovernment experts. In general company as well as public funds must go into the projects that are accepted. The public funds involved now are substantial. The percentage of industrial R&D financed by government in West Germany is not much lower than that in the United States, the United Kingdom, and France, despite that military R&D spending is much lower in West Germany.

Japan

For all the current hullabaloo, Western interest in Japanese industrial policies is of relatively recent origin.[11] Only in the late 1960s did politicians and scholars begin to take Japan seriously as a major industrial power capable of producing sophisticated products. Japanese textiles were one thing. But the ability of Japanese firms to take large shares of the American market for steel, televisions, and automobiles caused the United States to pay attention and ask what the sources of the Japanese miracle were.

Some economists writing on that question proposed that the development was not all that mysterious. Before World War II, Japan was a sophisticated industrial power and during the war demonstrated impressive technological capabilities. Although it came out of war destitute, since 1950 Japan had been able to achieve investment rates significantly higher than those of Germany and France and far higher than those of the United States and Britain. The educational attainments of the Japanese work force before World War II were close to European standards. After World War II the Japanese educational mill ground on at a furious rate and, by the middle 1970s, was turning out significantly more engineers per capita than the United States or the major European countries. From this point of view the miracle translates into very high rates of investment and of physical and human capital. The question then becomes how the Japanese are able to sustain these high rates.

Other scholars turned their attention to peculiarities of Japanese culture and institutions. Lifetime employment and its alleged implications was a trendy topic a few years ago. Recently it has been the Japanese style of management.

Interest in Japanese industrial policies, and Ministry of International Trade and Industry (MITI) in particular, is a Johnny-come-lately. I say this both to warn that, while recent scholarship is clearing up the matter somewhat, there still is some question exactly how Japanese industrial policies work and to note that these industrial policies are only one of several features that distinguish Japan from the United States and from the European nations.

The active, shaping role of the Japanese government in industrial development is not new. It goes back to the Meiji restoration of 1867, which was, after all, triggered by the shock of awareness of Japan's great technological and economic inferiority compared with Western development. Since that time Japan has been catching up. By the advent of World War II Japan clearly was highly competent in most of the industries that mattered for military production, a fact that Americans strangely seem to forget in talking about the Japanese postwar miracle. The instruments used after World War II were effective in the prewar era. The postwar MITI has recognizable connections with the agency that ran the Japanese economy during the late 1930s and through World War II. The current broad industrial policies of Japan have a long history.

The post–World War II era is different, however, in that the earlier era of Japanese industrial development was driven largely by the desire to achieve a strong independent military capability. Since the war Japanese industrial policies have focused almost completely on economic ends, although Japan has gradually developed along the way an ability to design and produce aircraft, rockets, and the associated electronics. In this way Japan is quite like Germany. After the war, when Germany dropped its military ambitions, however, it also dropped its directive industrial policies. Japan abandoned the former but not the latter.

Unlike the French, the Japanese appear never to have been fond of detailed quantitative targets for investment and output for particular industries. But the Japanese have taken seriously broad visions promulgated by MITI about the directions Japanese growth ought to take and even about the specific industries that ought to be stressed. Various instruments have helped that vision take concrete shape. In the early postwar years, MITI controlled access to foreign exchange and used this control both to keep foreign products out of markets in which it wanted to encourage Japanese industry and to determine which Japanese industries could import machinery and intermediate and raw materials. Detailed import licensing was gradually abandoned during the 1960s after Japan joined GATT, but MITI has retained power to keep out foreign goods in selected fields and has used that power. In the last few years formal controls have, however, largely dissolved. Protection now is to a large extent through custom and culture. MITI also has had the authority to keep foreign firms from establishing branches in Japan; and, while policies have liberalized over the years, by and large until recently foreign firms have been excluded from industries MITI has judged strategic. Here, too, recently the legal barriers have largely tumbled.

The Ministry of Finance in Japan long has had policies that restrict the ability of Japanese banks and other financial institutions to send funds abroad. Also the equity market is much less well developed in Japan than it is in the United States. Most of the large private savings in Japan thus flow to Japanese banks or insurance companies where they form a pool reserved for Japanese industries. The banks pay low rates to savers, lending rates are low, and credit is rationed.

As Zysman (1983) and Flaherty and Itami (forthcoming) have pointed out, this financial system is ideally suited for government guidance of investment. The leverage is exerted in part through government lending institutions but mostly through MITI guidance of private bank lenders. MITI in some cases has effectively exerted quite detailed control over the timing and allocation of new physical investments in an industry. Like the rest of the government control apparatus, the role of special finance in Japanese industrial investment now is quite small.

MITI has played an important role in helping the Japanese learn about Western technologies and manufacturing methods. Scientists, technicians, and managers have been sent abroad to observe and sometimes to study. MITI has regulated and channeled the flow of technology licenses. And over the last decade or so MITI has provided both R&D support and a mechanism for coordinating R&D allocation decisions in the high-technology industries it is pushing. Perhaps more important than any particular instrument has been the general agreement among the Japanese, including Japanese businessmen, that government leadership is not only legitimate but desirable and even necessary if Japan is to prosper, although there is occasional strong resistance.

In the late 1960s and early 1970s MITI began to put forth a vision of the Japanese economic future that placed heavy emphasis on the knowledge-intensive industries. The new vision forecasted a gradual shift in industrial emphasis away from shipbuilding, steel, and automobiles—which had been the industries stressed during the 1960s—and into consumer electronics, semiconductors, computers, and telecommunications. Japanese prowess in consumer electronics was already present and visible at that time. The policies in support of high-technology industries have involved the same blend of instruments used to further industries in the earlier era—initial protection of the home market, exclusion of foreign firms from Japan, assistance in learning about and gaining access to foreign technologies, favored access to credit, some efforts to mold the structure of the Japanese industry in a manner better suited to MITI's likings, and attempts to influence investments to take advantage of opportunities for cooperation and to avoid wasteful duplication. What seems special about Japanese policies toward high-technology industries is that MITI has played an active role in funding and orchestrating various large-scale cooperative research efforts aimed at helping the Japanese firms reach and then surpass foreign technological capabilities. I describe these programs in chapter 5.

5
Electronics, Aviation, and Nuclear Power

The foregoing discussion of government policies was broad and sketchy as seems necessary for a general description of a country's policies. The presentation of more detail inevitably reveals that policies have varied from sector to sector and even from program to program. Relatively detailed case studies of industries and programs, therefore, seem useful in providing another view of industrial policies.

Case study evidence has the liability of being piecemeal, scattered, and perhaps not representative. Also, in my view at least, only a few of the available case studies present enough detail so that one is confident that the picture being drawn is tolerably reliable. The advantage of good case studies is that they show more detail, so that one can begin to assess what the particular policies actually were and the effect they had. Where detailed studies exist of different national policies in the same industry, one can begin to hazard analysis of what works, what does not work, and why it does or does not work.

There are available case studies, of uneven detail and reliability, of the American, European, and Japanese experiences in semiconductors and computers, aircraft, and nuclear power. Continuing in the spirit of comparative analysis, I will attempt to sketch the similarities and differences in these experiences.

Semiconductors and Computers

The U.S. Experience. The U.S. semiconductor and computer industries, still clearly the strongest in the world, were enormously helped in their early days by a Department of Defense interest in the underlying technologies. While the details differ, the broad stories in the two industries are similar.[12]

Almost all of the exploratory R&D efforts that led to the early electronic computers were financed by the armed forces. The government was practically the sole market for the early operational computers and continued to be the dominant market into the early 1960s. Governmental funding of R&D and procurement were motivated strictly by national security interests. There is no hint that anybody in government thought that he was creating an industry that would be a major economic asset. Few of the companies involved in the early work for government believed that there would be a large civilian as well as a government market. Of course later a very large nongovernmental market for computers developed. The massive government support to computer technology provided U.S. companies with a head start that still has not been surpassed by foreign companies.

The U.S. experience with semiconductors has some similar elements and some differences. Perhaps the key difference is that the bulk of the early R&D was privately, not publicly, financed. The work leading to the transistor was motivated by perceptions of the utility of such a device for the telephone system. The rather special circumstance of AT&T ought to be noted. Bell Laboratories, the locus of the transistor invention, had long been noted as the world's premier industrial electronics research laboratory. In part because of concern that strength in research might enable AT&T to move beyond its near monopoly of the U.S. telephone system to a dominant position in other electronics-related fields, at the time of the transistor invention the company was under the cloud of antitrust prosecution. Not long after the transistor invention, a consent decree was worked out. Under its terms, AT&T agreed not to stray outside the field of telephones (except under government contract) and to license its technologies to other companies for use in other fields. It is unclear whether without the antitrust threat, AT&T would have gone into commercial production of transistors. Clearly, given the decree and the importance of the new technology to the telephone system, AT&T had not much to lose and much to gain by opening up the new technology to other companies so that they could help advance it.

The Department of Defense (DOD) quickly understood the potential of transistors for military hardware. Some R&D support was provided, but, perhaps even more important, the Department of Defense was clearly ready to buy transistors and systems designed around them if these could meet military needs. Miniaturization of electrical circuits was an important goal. The particular R&D projects financed by the government aimed toward meeting this need, however, turned out to be failures. The work that led to integrated circuits was not directly financed by the government. That work, however, was undertaken with the clear understanding that, if it were successful, there would be a massive government market. As with computers, government support was motivated by an interest in procurement, not an interest in establishing a national economic asset. Yet, also as with computers, such an establishment was one of the results.

It is important to note that the U.S. Department of Defense and NASA stood ready to buy semiconductors from any firm that provided a superior design. The key integrated circuit innovation and the development of the planar process for making integrated circuits came not from firms that had a long track record in electronics but from firms that were quite new to the game. Before the integrated circuit, DOD interest in semiconductors, although strong, was largely in anticipation of the advantages that improved semiconductors could lend. When the integrated circuit became available, both the Department of Defense and NASA made critical decisions to procure electronics equipment based on the new technology. The new firms were in the forefront first in the military- and space-procurement market and then in the civilian market for semiconductors that soon arose.

These American defense and space programs were massive compared with European and Japanese public expenditures on R&D in these industries and were far more ambitious in terms of the technological advances sought than anything tried by other countries. Before World War II American industry certainly was not laggard in electronics, but it was not noticeably superior to British industry; and German firms were considered the technological leaders. Several European firms were quick to develop transistors and, until the integrated circuit era, did not lag greatly behind American firms. But by the early 1960s, largely as a result of these defense and space programs, U.S. firms were the acknowledged technological leaders in computers and integrated circuits.

In the eyes of some observers, after 1960 the lead in computers and the lead in semiconductors went hand in hand (see in particular Malerba, 1983). The leading American computer companies increasingly provided the key market for advanced semiconductors. In turn American strength in semiconductors supported our computer lead. By the mid-1960s the civilian computer market was probably exerting the dominant pull for new technology, with military and space markets less important than before.

Several observers have questioned whether defense and space R&D programs still have the potential for pulling civilian technology in their wake. Thus executives of several semiconductor companies remarked, when the Department of Defense's recent Very High Speed Integrated Circuit (VHSIC) program was mounted, that the program would likely divert resources from the kinds of effort needed to keep U.S. firms in the technological forefront of commercial, principally computer-related, markets. Others have argued that the fear is misplaced, and that the VHSIC program is stretching the state of the art sufficiently in broadly relevant directions, so involvement is likely to help a company in commercial markets as well as in the defense market.

Recently there has been an interesting turn in the discussion. DOD officials responsible for the VHSIC program and for certain new programs aimed at enhancing computer design capability have indicated that they view these programs as directed toward enabling American companies to get or stay ahead of foreign firms in technologies of general importance. The explicit tie to defense procurement interests has been loosely drawn. In other words, DOD policies in this area have come to look like industrial policies rather than strictly procurement policies. I shall return to this discussion of the role of the Department of Defense later in this paper.

Britain. Although the funds have been modest and the ambitions restrained compared with the United States, Britain has invested nontrivial amounts of public funds in procurement-related R&D in computers and semiconductors.[13] Britain also has funded R&D with the express objective of boosting commercial competence of these industries. Despite the rhetorical objective, explicit assessment of commercial promise apparently has played little role in the allocation of these funds. The British government also has attempted to rationalize its computer and semiconductor industries as enhancing competitive capabilities. International Computer Limited was formed under government guidance. In the late 1970s the National Enterprise Board helped establish and support INMOS, a new company specializing in integrated circuits and oriented toward commercial markets. Total public R&D support, however, has been tiny compared with U.S. funding under defense and space auspices. British-owned firms have not been effective in generating exports and even in the home market have been relegated to niches.

France. The French have been much more aggressive than the British about building commercial competence. Their programs in support of computers and semiconductors, however, have been marked by dual purposes.

As noted earlier, French interest in developing a national capability to produce computers was motivated initially by restrictions imposed by the U.S. government cutting off access to an American computer judged necessary for the French nuclear program. The response was to establish a new national champion company in computers—Compagnie Internationale Informatique (CII)—and to mount a program of R&D support. Somewhat later, the French government also established a national champion for semiconductors. Significant R&D funding was provided under a series of programs. These moves marked a desire to build a French capability both to meet the needs of military procurement and to compete effectively on commercial markets. Zysman (1977) has argued that this built-in schizophrenia virtually guaranteed failure to achieve the latter objective. As with the British, French military R&D spending was not large enough nor were the objectives ambitious enough to pull the technologies beyond those of the Americans. At the same time, the military objectives and French pride required that French companies try to match the Americans where the latter were strongest. Public R&D support programs, allegedly commercial as well as military, have been quite directive. And company-proposed projects have been judged on the basis of how they fit government, not necessarily commercial, objectives. Thus the French companies could not hunt for commercial niches which could be developed into areas of major commercial strength.

While Zysman does not stress the fact, it is clear that the French would have liked to develop their industry by providing a protected home civilian market as well as a procurement market. And they have tried; general protection has been a hallmark of French policy in support of its electronics industry. The French interest in developing a uniquely French industry has been stymied by two factors, however. The first is that the incentives built into the French programs have led to some major tensions. As a prominent instance, CII, the subsidized computer company, resisted buying semiconductors from Sescosem, the subsidized semiconductor company, and bought rather from American firms that were producing more advanced products. Similarly, the French telecommunications companies had incentives to buy their supplies not from French companies but from American. Second, there always has been a problem about what being a "French" company meant. Recall the episode regarding Machines Bull. More generally, the strongest computer and semiconductor firms in France have been branches of American and Dutch companies. The French have found this situation extremely frustrating. During the 1970s French policy shifted from trying to establish a strong strictly French semiconductor and computer industry to encouraging joint ventures with American firms. Not many American firms would play the game, at least not under French rules.

Under Mitterrand, policy has shifted again toward a stricter nationalism. Pressures have been placed on certain branch firms to sell out to French-owned companies. For example, the French blocked imports of Japanese video cassettes and then arranged a joint venture of French and Japanese firms to produce in France. Clearly, however, the need to have Japanese participation is regarded as a thorn.

West Germany. The Zysman proposition that French policy has foundered in part because it has mixed military and commercial objectives is given some support if one contrasts the German experience. As I have noted, in recent years the German government has poured significant R&D funds into the German semiconductor and computer industries. The objectives behind these programs have been self-consciously commercial. While military R&D spending has increased significantly in recent years, the military and commercially oriented programs have been kept separate administratively.

The principal funding agency has been BMFT. BMFT has established several broadly defined areas for support; but, in contrast with the French situation in which the government has targeted particular firms and imposed strong pressures influencing the details of R&D allocation, as described by Friebe (1984), the German program works largely through the solicitation of proposals from companies. In recent years, Germany has made conscious effort to provide support for several companies, including small and medium-sized ones. In general BMFT supports less than half the cost of an R&D project. Although the German industry has not achieved outstanding success in the market for either semiconductors or computers, these German industries are recognized as being significantly stronger than the French. The major German commercial successes have been not in those fields that directly confront the most advanced American products but rather in producer goods electronics. Philips, the Dutch-based international firm, also has done reasonably well, at least up to recently in a niche—consumer goods electronics.

Until recently, almost all the R&D support and promotion of their electronics companies by European governments has been at a national level. As in the earlier aircraft promotion, over the past few years sentiment in Europe for joining forces has been rising. The new program of the European community, the European Strategic Program for Research and Development on Information Technologies (ESPRIT), has a structure that has elements in common with Japan's MITI programs.

Japan. Like the German and unlike the French, Japanese policies have been shaped not by an interest in the ability to produce weaponry but rather by the desire to establish a commercially profitable industry.[14] Since Japanese success in electronics is perhaps the most often cited example of successful government policies in support of high-technology industries, it is worthwhile to discuss this experience in some detail.

The rapid takeover of the American color television market by Japanese manufacturers in the late 1960s came as a shock to many Americans, and it was, rightly, widely regarded as an indicator that American preeminence in consumer electronics was threatened. This episode, however, followed earlier Japanese successes in capturing a large share of the American market for transistor radios and black and white television sets. The data show that by 1960 Japan was employing many more semiconductors than any European country, including France and Britain, despite the absence of any major military procurement program. So, when the Japanese began to develop their color television industry, they did so from a base of considerable experience in consumer goods electronics. By far the largest market for Japanese-made television sets was the protected home market, and the earlier Japanese sets were designed with that market in mind. It turned out that there was also a large U.S. market for small color television sets, which American companies were not producing. Japanese color television exports to the United States began by hitting that market.

What role did explicit industrial policies play in this development? Certainly they provided broad encouragement, protection of the Japanese home market, and the standard Japanese assistance for exports. In addition, MITI helped fund a cooperative research program that enabled Japanese television producers to surpass American companies in fully exploiting the opportunities afforded by integrated circuits. This support was for generic research, not for the design and development of specific products. The Japanese companies themselves initiated and funded product design and development. Sony's work, which led to its special tube design, was not funded or even encouraged by MITI. Peck and Wilson (1982) have remarked that color television was not an industry targeted by MITI. But the issue is a matter of degree not kind: MITI certainly encouraged and aided the industry.

Japanese policies supporting their semiconductor industry have a similar flavor. The large protected home market for semiconductors was supplemented by policies of government-controlled enterprises, particularly Nippon Telephone and Telegraph (NTT), to procure equipment that used Japanese-made semiconductors. The role of NTT in the rise of Japanese competence in semiconductors is viewed by several scholars as much more important than that of MITI. NTT, like AT&T, has a powerful set of research laboratories, a counterpart to Bell Labs; but, unlike AT&T, NTT has no in-house facilities for product engineering and production. In place of a Western Electric counterpart, the NTT laboratories have forged intimate relations with the R&D and production facilities of several of Japan's most important electrical equipment producers. Thus NTT's procurement-related research, and its funded design and development projects located in the facilities of its suppliers or in cooperative facilities, enhanced the commercial competence of those suppliers in a way that AT&T procurement, tied to Western Electric, could not enhance the commercial capabilities of the U.S. semiconductor industry.

The same consent decree that reinforced AT&T's willingness to open transistor technology to other companies blocked Western Electric from exploiting on the commercial market what it learned in later telephone-connected semiconductor R&D. While Bell Labs and Western Electric patents continued to be available for license, one learns far less from a license than from actually doing the work in the first instance. And the patents owned by AT&T were available to everyone, not just to American firms.

The so-called VLSI (very large scale integration) effort of the middle 1970s involved both a program by NTT designed to develop and ultimately procure integrated circuits suitable for telecommunication uses and a more broadly oriented program sponsored by MITI to bring Japanese companies to the forefront of semiconductor technology relevant to computers. Although the latter has been more widely publicized, it involved much lower funding levels. Perhaps the most striking feature of the MITI program was that it was organized around several corporative research laboratories, staffed by scientists and engineers drawn from the involved companies, with the funding shared between the companies and MITI. This program, as the earlier one directed toward color television technology, was largely generic in nature. While many patents came from that program, the basic purpose and result of the program was to bring Japanese companies up to present technological development along a rather wide front. Although MITI did not attempt to push particular commercial product developments, however, the projects were carefully chosen for their likely commercial relevance. Companies whose personnel engaged in a particular successful project had a definite advantage in commercial design.

The involved companies felt this imbalance very much. The result was, on the one hand, restrictions on the program to avoid areas in which particular companies already had a proprietary interest and, on the other, jealousies among the companies regarding the projects they were assigned. Apparently strong and subtle leadership was needed to hold the program together. Analysts diverge on how important they think the program was in bringing Japanese semiconductor capability up to the frontiers. Certainly the funds were small in relation to those involved in the in-house efforts of the Japanese firms or in NTT-financed work. But some observers regard the program as having played an important catalytic role.

The case of computers is somewhat special because of the presence in Japan of IBM. IBM entered Japan before World War II and its leverage on the Japanese also was enhanced because it held some of the basic computer patents. MITI successfully bargained with IBM for licenses and got IBM to limit its Japanese sales, but IBM remained the largest computer company in Japan until 1981 when it was surpassed by Fujitsu. To help offset IBM's advantage, MITI helped the Japanese computer companies establish a computer-leasing company so that like IBM they could offer their machines on lease. Japanese government purchases of computers have virtually all been from Japanese firms.

In the late 1960s MITI apparently made a judgment that Japanese computer capability was too fragmented and that merging would be in order. The large Japanese electronics companies proved unwilling to separate out their computer design and manufacturing capabilities and to merge these. MITI had similar trouble earlier when it tried to rationalize the Japanese auto industry. As a compromise, MITI organized and helped support several research and development groups, each group oriented around a particular strategy for computer design and commercialization. The target for these efforts was not a government market that could be ensured and shared by the cooperating firms but the highly competitive general commercial market. Because of this target, the cooperative R&D arrangements often proved fractious since the work being done touched on the potential proprietary interests of rival firms. More recent Japanese programs have stressed basic and generic research. Unlike the earlier program, the fourth and fifth generation computer programs apparently do not involve particular companies in commitments regarding the nature of the computers they ultimately will design and market. Also unlike the earlier programs, there has been less insistence that the companies contribute some of their own funds. The requirement that companies repay funds if there are financial successes has been virtually abandoned, reflecting the abstruse and generic nature of the research.

Peck (1983) notes the comprehensiveness of the fourth and fifth generation programs. They clearly are designed to develop the abilities of the major Japanese computer manufacturers to move in many possible directions as the technologies develop and the nature of the markets becomes clearer. As with the earlier MITI R&D support programs, the public money involved is very small compared, say, with the funds the Department of Defense put into the U.S. industry in the 1950s and 1960s. The funds are small compared with the proprietary research funded by the Japanese computer companies. What MITI appears to be trying to do is not to direct the commercial development of computers in Japan but to see that the Japanese companies have the technological capabilities to compete with IBM and the other major Western companies in designing and developing the next generation of computers.

As earlier in the United States, a dynamic computer industry increasingly provides a market-inducing technological advance in semiconductors, principally integrated circuits. Unlike French manufacturers, Japanese computer manufacturers buy largely Japanese-made integrated circuits. This situation certainly is due partly to the fact that the large computer manufacturers are also the producers of semiconductors, but it also is the result of strong MITI urging.

Aircraft

The story of government policies in support of civil aviation contains a number of elements in common with the electronics story.[15] To a far greater extent than in electronics, however, governments, particularly the British and French, have financed the development and subsidized the production of particular designs aimed explicitly for a civilian market.

The U.S. Experience. Except for the supersonic transport, the U.S. government has been unique among the five countries considered here in not involving itself in deliberate direct subsidization of civil aircraft development. During the period between World War I and World War II the government took a direct interest in the development of the U.S. aircraft industry. The National Advisory Committee on Aeronautics was established in 1915 to "investigate the scientific problems involved in flight and give advice to the military air services and other aviation services of the government," (Mowery and Rosenberg, 1982). As the statement of mission attests, the program was justified in terms of direct government (largely military) needs; but, from the beginning, the problems NACA worked on were common to commercial as well as to miliary aircraft. NACA's work on engine and airframe streamlining played an important role in enabling the design of the Douglas DC-3. That aircraft and the planes that evolved from it

(DC-4, DC-6, and DC-7) dominated the commercial airliner market from the mid-1930s until the advent of passenger jet aircraft. During this pre-World War II period, the government subsidized the airlines and indirectly, therefore, civil aircraft design and development through contracts to carry airmail.

By the late 1930s NACA began to concentrate more specifically on problems of special interest to the military, and the flow of civilian benefits diminished. After World War II, much of the generic research mission that NACA had shouldered was shifted to the aircraft companies through DOD contracts explicitly with those companies. By the late 1950s NACA had been transformed into NASA, and the orientation shifted largely toward space.

Although technology relevant to military aircraft and that relevant to commercial aircraft always have differed in important respects, until 1970 or so there was considerable overlap. During the post-World War II era, design and procurement of a new aircraft or a new engine for military use often led technological advance with civil technology following. As noted, the American postwar preeminence in the commercial aircraft business arose directly out of military research and development and procurement contracts. The Boeing 707 was designed by the same company at the same time with a plane bought by the Air Force that had many design elements in common. The American wide-bodied jets show their origins in military cargo planes and the engines that powered them. Until the supersonic transport episode, there were no programs of the U.S. government meant expressly to help in the development of commercial airliners, nor was there any pressure for such programs from the major aircraft producers.

Europe. The situation in Britain and France has been quite different from that in the United States. In Britain, during World War II, a relatively explicit government plan was drawn up for postwar support of the design, development, and production of civil aircraft. During the early postwar years several subsidized designs were developed according to the plan. Most of these efforts were aborted before a vehicle was ready for a market test. The few designs that were fully developed turned out to be in areas dominated by American aircraft.

The British plane that marked the largest technological step forward, the De Havilland Comet, which was the first commercial turbojet transport, was developed and produced without government support. Turbojet aircraft were not in the plan. The Comet, which was produced and used six years before the Boeing 707 and the French-built Caravelle, had fatal technical problems. Metal fatigue caused several disastrous crashes. Government funds did go into efforts at redesign, but the funds were not sufficient to effect the needed modifications in time to beat out Boeing.

The experience of the British government of betting right was no better during the 1950s and 1960s than it was in the immediate postwar period. During this time the government subsidized the design of more than a dozen aircraft. Only one—the Viscount—can be regarded as close to a commercial success. The nationalized British airlines, BEA and BOAC, were coerced into buying British planes and, as a result, often were disadvantaged in relation to other airlines that flew competitive routes and had freedom to shop.

The British government involved itself in selecting and financing aircraft development projects and in deciding what planes the British airlines should buy. During the 1960s it also pressured a reorganization of the British airframe- and engine-manufacturing industries and through mergers reduced significantly the number of independent companies. The government hoped thereby not only to exploit economies of scale better but also to reduce pressures for government sponsorship of many projects to maintain company employment. At the same time the government changed its method of financing and became a formal business partner in the development of aircraft, expecting to share in the profits as well as in the costs. As noted, there were no profits to share. And the losses of the airlines and of the companies had to be picked up by the Treasury.

In the middle 1960s, partly in response to the financial losses, a committee headed by Lord Plowden was formed to consider the future of the British aviation industry. One of the committee's most important recommendations was that efforts should focus on collaboration with other European countries. It was already clear that one ongoing such effort—the Anglo-French Concorde—would likely be a financial albatross. The logic of the Plowden recommendation, however, seems to have persuaded the British government that attempts to develop a purely national industry through subsidization and a guaranteed home market were extremely expensive and ultimately futile. The recommendation foreshadowed several cooperative ventures during the 1970s, notably Airbus.

The French story has some similar and some different aspects. France has had subsidy and government direction of civil aircraft development and a built-in home market in the airlines, but the French effort has been less scattered and, by-and-large, more successful. During the 1950s the French government authorized the development of the turbojet Caravelle. The plane, designed for short- and medium-range trips, found a niche in the first-generation jet market because the other planes—the 707, the DC-8, and the Comet—were designed for longer range travel. The Caravelle, however, was surpassed by the Boeing 727 which appeared in the early 1960s.

Except for the Caravelle, during the 1950s the French government did not really push or try to direct commercial aircraft design and development. Efforts were focused on military aircraft. There appears to have been little of the urgency to establish or preserve a commercial aircraft industry that marked the British experience, perhaps because during the war Britain had built up a large labor force in its aircraft industry and France, of course, did not.

France's next major venture in civil aviation was the supersonic aircraft, the Concorde, a joint venture with the British begun in 1962. The French interest in the venture flowed from deliberations as to the appropriate successor to the Caravelle, which was by then obsolete. The British, frustrated by their experience in developing and producing a long-range plane directly competitive with American planes, were interested in a technically advanced transoceanic plane.

Enough has been written about the Concorde so that only a sketch is required here. In contrast with Airbus planners, those planning the Concorde paid very little attention to the nature and size of potential markets or to the sensitivity of those markets to price. Nor did they heed the experience in military R&D that the cost of ventures aiming for a radical advance in technology tends to be greatly underestimated. The original $450-million estimate for development costs proved low by a factor of ten. Only the captive French and British airlines could be forced to accept delivery of the Concorde when it was finally ready for commercial operation in 1976, and both governments have had to subsidize the operation of the plane. Production was terminated in 1979; only sixteen aircraft had been produced.

The U.S. government also was drawn, or it jumped, into subsidizing and directing a supersonic transport project. The U.S. effort, begun several years after the European effort was launched, was a direct response to it as well as a desire to exploit expected spillover from the development of the B-70 strategic bomber prototype. The normal procedure in the development of specifications for a new commercial aircraft, in which there is significant interaction between the airlines and the company considering the venture was not followed. Instead the lead government agency, the Federal Aviation Administration, stipulated the performance requirements, with little consultation with the airlines. Boeing won the contract. Serious technical problems (the original design proved infeasible), cost escalation, and opposition from environmental groups led to the program's demise in 1971. The experience with Concorde suggests the United States was lucky that the program never achieved a technically viable aircraft.

The Airbus is an entirely different story and, since not much has been written on it, warrants telling in some detail. As early as 1963 Britain and France had discussed a possible joint venture to produce a large commercial subsonic aircraft. By the mid-1960s the Germans, who were eager to expand their presence in the aviation industry, joined the discussions. The German aircraft industry had been dismantled after World War II. During the 1960s, with encouragement by the U.S. government, German companies began to produce the Lockheed F-104 fighter under license. Although there were also a few small commercial endeavors, nothing major was attempted before Airbus.

France, Britain, and Germany reached a rough agreement to develop an airbus by the fall of 1967. As Lorell (1980) tells the story, the early days of the venture were marked by considerable intergovernmental squabbling, particularly between Britain and France, regarding both the details of the design and the division of responsibilities. For various reasons, including discontent with the way the project was taking shape, in 1969 the British government removed its support. After that time, although the international pulling and bargaining did not cease, French ideas about what the project should be and how it should be run gained and maintained ascendancy. While German financial support was close to that provided by France, the leadership and the direction were largely French. And this ascendancy continued even after other participants—first the state-owned Spanish aircraft firm, then the British firm Hawker-Siddeley, and in 1979 the British government again—joined the group.

With the diminution of intergovernmental conflict about the nature of the program came a redefinition of goals that stressed even more the achievement of a commercially viable vehicle. By 1969 a decision was made to start developing a 250-passenger wide-bodied plane tailored to the relatively short runs of the intra-European passenger air network. The group chose this market niche in the course of considerable discussions with the European airlines regarding the kinds of planes they would like to procure. By this time the Douglas DC-10 and Lockheed L-1011 were well under development. Both were planes of roughly the size proposed for Airbus, but they aimed for longer flights. Relatedly, the Airbus was designed around two engines rather than three.

With the cutting back of the pork barrel aspects of the program, the governments involved in effect agreed not to meddle in the details. Under the accord that officially launched Airbus Industrie in December 1970, the top management of the involved firms were granted the authority to define both technical and marketing objectives for the project. Although the participating governments hold the purse strings and thus ultimately can veto decisions, government officials do not become directly involved in formulating design or marketing proposals. The top executives of the firms also have the authority over administration and thus control how the decisions are implemented. The contrast with Concorde or the SST program is dramatic.

Despite a design apparently well aimed for a market niche, (actually, two designs by the late 1970s) and despite a promising management system, during most of the 1970s the financial prospects for Airbus seemed dim. Through the late 1970s orders for Airbus were slim compared with those for the Lockheed and McDonnell-Douglas planes. In 1979 Airbus orders began to pick up dramatically. While it is still too early to tell if the consortium will make a profit, its planes have sold better than any other European-designed airliner ever made.

The fierce competition among the Airbus consortium, Lockheed with its L-1011, and McDonnell-Douglas with its DC-10, for roughly the same market reveals sharply the conflictive nature of national policies in support of high-technology industries for economic purposes. The American companies complained, naturally, that foreign governments were heavily subsidizing their competitor.

Japan. The Japanese aircraft industry, like the German, was dismantled after the war. The industry started operating again during the Korean war with production, under license, of several American military aircraft. Typically, Japanese coproduction started as assembly of U.S.-produced parts and gradually extended to encompass more basic production. Several small commercial aircraft projects were pursued in the 1960s. The most ambitious commercial venture was the YS-11 aircraft, designed and built by a consortium of Japanese firms MITI helped arrange and funded substantially by the Japanese government. The aircraft, designed for relatively short flight, was widely used on Japanese routes but did not sell well abroad.

During roughly the same period, several of the large Japanese firms began to take work as contractors, building certain components and assemblies, for the Boeing 747 project and for other new American aircraft. Under MITI's guidance a new consortium of firms formed to enable more comprehensive involvement of Japanese firms in aircraft development. In 1978 the consortium signed an agreement with Boeing that involved the consortium in considerable work in the fuselage of the 767 aircraft. MITI has provided half the funds, in the form of interest free loans, to be repaid only if the project turns a profit. A similar arrangement has been worked out for collaborative engine development and production between a Japanese consortium arranged through MITI and Rolls Royce. Recently, Pratt and Whitney joined the group.

In 1984 the Japanese aircraft consortium signed a new agreement with Boeing for the design and development of a wholly new 150-passenger aircraft that would use the engine currently under joint development. In this agreement, in contrast with the 767 agreement, the Japanese will have substantial design responsibility and will operate more as partners and less as subcontractors. This effort represents, therefore, a substantial step toward a significant Japanese presence in the world civil aircraft industry.

Several well-informed analysts believe that in the coming years Japan will be a powerful independent player in the commercial aircraft game. The obstacles to the creation of a strong Japanese aircraft industry, however, appear substantially greater than the obstacles in electronics. As Mowery and Rosenberg (1984) point out, the internal Japanese market is far too small to support the industry and thus to serve as an incubator. From the beginning of the venture, commercial viability depends on sales in a world market. Second, R&D funds are concentrated to a much greater degree on particular designs and products. Thus the Japanese government is, willy-nilly, drawn into making particular commercial bets, highly concentrated ones at that. Third, the Japanese are not entering the aircraft industry alone but as parts of various international consortia. Several of the foreign partners are technologically stronger than the Japanese and may be reluctant to teach the Japanese all that they know. The Japanese government, however, has not invested much in basic and generic research relevant to aircraft and aircraft engine design to provide the Japanese companies with an independent source of technological strength.

In an industry like aviation, in which international joint ventures are becoming the standard way of designing and producing new vehicles, one can ask what is a national capability and what is truly international? The growing internationalization of high-technology industries is a topic to which I shall return.

Power Reactors

In the field of nuclear power, the government of the United States, as well as the governments of the major European countries and of

Japan, has spent enormous sums of money over a long time to create a commercially viable and internationally competitive power reactor industry.[16] In each of these countries a special government agency has been charged explicitly with the job of guiding reactor development, and in several countries the agencies have done this in great detail. Although by some standards the French and Japanese programs might be regarded as reasonably successful and the German program potentially so, it is unclear that so far the rate of return on any of the programs has been positive.

The issues are complicated and tangled, however. First, even more than in the cases of aviation and electronics, policies in support of the development of nuclear power technologies have been tightly intertwined with explicit national security objectives, at least in the United States, Britain, and France. Second, in the early days of atomic power, concerns about environmental impact and safety were muted. As these concerns became better articulated and better represented in the political process, new design requirements and more stringent licensing requirements were imposed. The financial costs of nuclear power thus significantly increased. Further, at roughly the same time that these factors were slowing the tide of nuclear energy, economic hard times set in and forecasts of the growth of energy demand were drastically scaled down.

The entanglement with national security objectives made it more or less inevitable that a government body would exert detailed control of the development of the technology and that noncommercial values would be given a prominent place. The rising concerns about safety and the changes in perceived long-run economic prospects turned somewhat sour the initial high hopes about the economic advantages of nuclear power.

The United States. Shortly after World War II the American Atomic Energy Commission was established and was assigned responsibility for future nuclear developments, civilian as well as military. As the same time the Congressional Joint Committee on Atomic Energy was established. For the next quarter century the executive agency and the congressional committee worked closely together and, in effect, jointly reigned over the government programs in question.

The programs in support of civilian nuclear power grew out of the programs to design and develop nuclear power reactors for submarines and surface ships. President Eisenhower's "atoms for peace" speech in 1953 signaled and put in place a commitment of the U.S. government to develop civilian nuclear power reactors. The sense that it was important to get power reactors designed and built quickly, which marked the Eisenhower speech, also reflected the views of the Atomic Energy Commission. It meant that the bulk of attention focused on the light water reactors for which some experience had been accumulated in the naval programs. Light water reactors used enriched uranium as a fuel, but the United States had ample enrichment plant capacity, built in support of the nuclear weapons programs.

The major companies that entered the business of designing and producing reactors and the utilities were bullish about the prospects and invested significant amounts of their own money. The Price-Anderson Act of 1957 limited the liability of utilities in the case of nuclear accident. The Atomic Energy Commission supported research, offered some financial backing for experimental and demonstration plants, and, most important, urged and pushed the companies and the utilities to produce.

It was apparent from the outset that, if nuclear power were to be competitive with conventional power, the plants would have to be very large. Thus during the late 1950s the companies committed themselves to produce and the utilities to buy nuclear power plants very much larger than any that had been actually built and tested. In this era of optimism very little attention was paid to issues of reactor safety or to questions of waste disposal.

The Shippingport demonstration plant went into operation in 1958, followed by the Yankee Nuclear Power Plant in 1961. The Atomic Energy Commission subsidized both of these plants, which operated at scales far smaller than those the companies and the utilities already were committed to produce and use commercially. The objective was to gain experience from their design, construction, and use. The faith was that "scaling up" would pose no serious problems. In 1963 a contract was signed for the first full-scale reactor, judged competitive without subsidy.

As it turned out, the companies who contracted to build the reactors could not do so at costs anything close to the agreed-upon price. Also, the large-scale reactors had major technical problems that had not been apparent with the smaller demonstration versions. The first generation commercial reactors were not competitive with conventionally fueled power sources, and the companies who produced them lost money. The utilities that procured them undoubtedly could have produced electricity at lower cost had they build up-to-date conventional plants, despite the heavy front-end subsidy of the Atomic Energy Commission and subsidization of fuel costs.

During the 1960s, despite this unfortunate early experience, the companies continued to try to sell and utilities continued to order versions of the light water reactors. Disenchantment set in gradually. As noted, the concern about environmental effects and safety rose, and then, somewhat later, the projected growth of demand for electric power fell sharply. The large jump in oil prices and more optimistic beliefs about future availability of uranium in relation to demand by themselves made the nuclear power alternative look more attractive in relation to conventional plants. The sharp rise in estimated nuclear plant costs associated with new environmental and safety requirements and the now much more complicated and time-consuming regulatory process, however, deterred many utilities from taking the nuclear route. Aside from bringing into operation several plants whose construction started some time ago, nuclear power expansion in the United States has come to a virtual standstill.

In the early 1960s, on the belief that its first round of objectives had been achieved, the Atomic Energy Commission shifted its attention toward research and development on a breeder reactor. The case for the breeder reactor rested, in large part, on forecasts that there would be considerable growth during the last decades of the twentieth century in the number of regular nuclear plants and that supplies of uranium would therefore be consumed relatively quickly. As it did with conventional reactors, the Atomic Energy Commission early committed itself to a particular type of breeder reactor—the liquid metal fast breeder reactor. Considerable funds went into research and development on this reactor. By the middle 1970s, however, skepticism began to be voiced strongly. In the first place, projections of growing scarcity of uranium no longer seemed justified. Second, concern that breeder reactors generated materials that could be used in bombs intensified. Many studies showed that no economic case could be made for going ahead with at least this particular breeder reactor program. Nonetheless funds continued to go into the Clinch River breeder reactor project. Although the old Atomic Energy Commission had been dead for more than a decade, the political momentum of the projects it initiated proved hard to slow down. In late 1983, however, Congress stopped funding the program.

Britain and France. Although the stories of the British and French programs have some essential things in common with the American experience, they have some important differences. One major difference is the following. After the war both the British and the French opted for a gas-cooled graphite-moderated reactor design for two central reasons. First, because these reactors used natural uranium as a fuel, their employment in a power grid did not require access to enriched uranium, which in the early postwar era only the United

States could produce. Second, these reactors produced plutonium as a byproduct and thus were a natural part of a program meant to develop a military nuclear capability.

The British Atomic Energy Board, later the Atomic Energy Authority, has at least until recently exerted even more detailed control over the development of nuclear power than did the U.S. Atomic Energy Commission. From the beginning it has been committed to its own designs, which have basically stuck with the early commitments to gas cooling. Electric power generation and distribution in Britain is nationalized and centralized. The Central Electricity Generating Board was, after its early experiences with experimental plants, increasingly skeptical about the economic merits of gas-cooled reactors and over the years has pressed for light water reactors. A succession of committees has been charged to resolve conflicts between the AEA and the CEGB. In part because the AEA remained the principal source of technical expertise heard by the British government and in part because of a desire to stay with reactors designed and built by the British, until recently the conflicts have been resolved in favor of the AEA's designs. There has been an almost endless tinkering with the structure of the reactor industry in hope that reorganization there would resolve the increasingly obvious shortcomings of the plants placed on line.

Britain's reactors have not found a market abroad and have been employed domestically only because the Electricity Board has been, in effect, ordered to use them. In the late 1970s and the early 1980s this situation was reluctantly recognized at the top. The power of the Atomic Energy Authority to dictate the path of nuclear power development in Britain apparently has been attenuated.

Although the French situation has something in common with the British, from the beginning the authority responsible for the nationalized power network, Electricité de France (EdF), has been a more effective counterweight to the Atomic Authority than has been the case in Britain, and the French program shifted orientation significantly before the British did. As it gradually became more expert, EdF became skeptical about the economics of gas-cooled graphite-moderated reactors, just as had the British Central Electricity Generating Board. In the middle 1960s in France, as in Britain, the central government authorities ruled for the atomic energy authority and against the electricity authority in cases of conflict. EdF, however, was able to fund work on light water reactors itself and to keep the options open. By the early 1970s, with the passing of Charles de Gaulle, EdF began to win the upper hand and to gain authority regarding reactor development and purchase.

By the middle 1970s France had shifted almost completely to pressurized water reactors as the technology of choice for the short- and medium-run. To a greater extent than in the United States, the designs were standardized, and the French company engaged in the production of such plants, Framatone, began to get the advantages of economies of scale and experience. According to one study, although U.S. nuclear plants cannot produce electricity as cheaply as modern coal-fired ones, French plants can, at least given the high costs of French coal. As in the United States, questions have been raised about the environmental effects and the safety of reactors; however, the French government has been quite authoritarian in putting down protests. Although recognition that future demand for electricity will not be as great as forecast has slowed down construction, all new electricity generating capacity in France now is nuclear, and production is planned ahead at a modest rate. France continues to work, now increasingly in consort with other European countries, on a breeder reactor.

Germany. The German story diverges from the British and French. Again, that Germany was not trying to build a military capability is important to recognize. Also, Germany had no strong resistance to dependence on the United States for fuel. Given the questions explored in this essay, however, the most important difference is probably that the strong centralized control of reactor development that marked the British and French experience and, to a lesser extent, the U.S. experience never took shape in Germany.

For a period after World War II Germany was expressly prohibited from engaging in nuclear research activities, and only in the 1950s did the constraints loosen and the Ministry for Atomic Questions form. Historically, the Länder have had major responsibility for funding research at the universities, and, as Germany began to reestablish a nuclear research capability, that responsibility was not centralized as it was in other countries. Also, in Germany, like the United States and unlike France and Britain, electricity production and distribution is decentralized—there are several independent utilities—and cannot be directed from the capital. The larger German companies, principally Siemens and AEG, had been watching reactor developments for some time and when the German program began, had their own ideas about the most promising roads to follow.

The programs of the federal government, therefore, were, from the beginning, only a part of the venture. There were several different sources of initiative.

The Eltville program, initiated in 1957, had the express aim of helping German firms develop capabilities to do more than simply copy foreign (generally American) designs. The companies received subsidies to work on designs they, as well as the funding authorities, deemed promising. As Keck (1981) discusses, in the late 1950s and early 1960s the ministry attempted to lay out a more coherent plan, with priorities, and to take a more active role in allocating R&D resources. The major German firms proved willing to undertake projects proposed by the ministry so long as they did not have to put up any of their own funds, but the companies laid their own money on what they thought were the best bets. And at that time the companies were much less concerned than the ministry about the fact that they were basically simply learning to build American designs. The utilities also were more narrowly economically oriented than the ministry, and the signals they gave to the companies reinforced inclinations to proceed relatively conservatively.

Keck's story of the German fast breeder reactor program brings out especially sharply the difficulties with the ministry's program. By the mid-1960s scientists at the ministry-financed Nuclear Research Center at Karlsruhe became convinced that unless a strong West German program were mounted quickly, the German nuclear industry would be greatly handicapped in relation to the American in the design and production of breeder reactors. These German scientists believed breeder reactors would become the dominant technology by the middle or late 1970s. Various consortia for German companies were induced to work on several designs pushed by Karlsruhe, but these companies did not put any of their own resources into work on breeder reactors, which they themselves felt were not likely to be commercially viable for some time. The companies' beliefs on the matter surfaced only when the ministry requested that company funds go into the projects that the firms were unwilling to do.

The companies' funds were going into design and development of technologies the companies thought would be commercially profitable; in the early stages these technologies were generally designs that American firms advanced. By the late 1960s German companies had acquired sufficient competence to cut their ties with American firms. German reactors were competitive in world trade.

The experience of the German industry following the oil shock, however, has been more akin to that of the American than to that of the French industry. As in the United States, citizen group opposition to nuclear power has become quite strident and has not been squelched. Falling expectations about future energy demand have cut back on orders. The prospects are unclear.

Japan. The Japanese case is marked by fewer sharp turns and obvious technological mistakes than are revealed by the histories in the other countries; but Japan, too, currently is experiencing citizen resistance to a technology widely regarded as oversold and dangerous. Also, the Japanese case, as the others, clearly reveals the entangling of the reactor development programs with international politics, although the tangle was not of Japan's making.

Less than a decade after Hiroshima and Nagasaki, government and industry leaders in Japan, encouraged by the Americans, began to plan for the development of nuclear power. After showing a brief interest in British gas-cooled reactor designs, the Japanese fastened on American technology and adopted the American long-range plan for nuclear power development. This plan meant light water reactors for the short and medium run, with an accompanying commitment to obtain enriched uranium, to increase the use of fuel reprocessing, and ultimately to adopt a breeder reactor. This strategy has been worked out and implemented in Japan through the close cooperation of several industrial and governmental bodies. The key actors have been the major regional electric power companies, the companies that design and produce the reactors and their components, the science and technology agency which has had main responsibility for managing nuclear R&D effort, and the Japan Atomic Energy Commission. MITI's role has been mainly that of licensing, regulating safety, and inspecting plants. Since 1978 a Nuclear Safety Commission has also existed.

As in the other countries, government-provided funds have accounted for the main share of nuclear basic and generic research and experimental development. The companies and the utilities have paid for the production and implementation of designs that are regarded as relatively well worked out. The Japanese producers, as the German, quickly mastered American light water technology. By 1980 Japan was second only to the United States in the amount of nuclear power on-line.

From the beginning of the program, a key Japanese objective was to cut back on requirements for imported petroleum and for high-cost domestic coal. The oil shocks of the 1970s strongly reinforced this objective. A recent study reports that in Japan the cost of producing power with nuclear reactors is less than the cost of using coal-fired plants. The strikingly low cost of capital and the high cost of coal in Japan must be important factors in that calculation. By some standards, however, the Japanese program looks successful.

Japan faced, and still faces, two major problems regarding nuclear power. First, the Japanese decision to use light water reactors in the early days made Japan dependent upon American providers of enriched uranium. The adoption by the Japanese of the broad American strategy of nuclear reactor development led them into developing enrichment capacity as well as the capacity to recycle spent fuel elements and into a commitment to the breeder reactor as the technology of choice for later in the century. As noted earlier, under the Ford administration the United States backed away from this strategy on grounds both of changing beliefs about the future demand and supply for uranium and of concerns about nuclear proliferation; it also exerted considerable pressure on other countries to abandon plans for reprocessing. Although Japan has not bowed to this pressure, the situation has been uncomfortable on several occasions.

Japan also has seen rising citizen resistance to locating nuclear plants near population centers, which greatly narrows the available sites. As in the United States, citizen concerns involve both environmental issues—in particular, citizens fear that shore-based reactors will hurt Japanese fisheries—and safety issues. The nuclear accident that occurred in one of Japan's reactors in 1981 has highlighted safety problems, and in Japan, as in the United States, gaining agreement about a plant's site and design, its construction, and its operation is now a time-consuming and costly business.

6
What Lessons?

I concluded chapter 1 by raising several broad questions. Are leading industries "strategic" and, if so, in what sense? Does general economic and technological strength or do special policies enable a nation to have prowess in leading industries? If the latter, what kinds of specific policies are important? We now have the basis for hazarding answers to these questions. Earlier I stressed that the way I characterize technological progress and what I choose to describe about government policies are very much influenced by my theoretical preconceptions. It is even more evident that the way I interpret the record and the tentative answers I provide to the basic questions come from my mind's eye and not simply from objective observation.

General Strength or Special Policies?

The answer to the question of whether it is general strength or special policies that lead to national capability in high-technology industries is probably both. I read the record as indicating that general strength is a necessary condition. Given basic technological and economic strength, however, the right policies specially aimed at the high-technology industries certainly seem to have lent advantage to national firms.

It is hard to escape the conclusion that general strength in scientific and engineering education and research is a prerequisite for strength in high-technology industries. The technological preeminence of the United States in these industries since World War II surely has something to do with the fact that, although in recent years our educational advantages have diminished, we still have a larger ratio of scientists and engineers to the total work force than any other country in the non-Communist world. Since the late nineteenth century, Germany has been noted for the quality of its scientific and technical education and the skills of its work force from scientist and engineer to technician and mechanic. Japan's rapid surge toward the frontiers clearly has been associated with the remarkably large fraction of its population that has been getting a technical education. Britain's decline in relation to Germany and the United States, and recently in relation to France and Japan, has been attributed at least in part to weaknesses in the British educational structure.

I read the evidence as suggesting that the key is a system of scientific and technical education that both trains well and points a good percentage of graduates toward industrial careers, not necessarily preeminence in academic science. Of course technical education and academic science are not disconnected. It is almost impossible to train high-level scientists and engineers for work in industry unless one has a university faculty operating at or close to the frontiers of knowledge in their fields. Britain, however, has stayed in the forefront of the relevant academic sciences but has not managed to establish a culture wherein a significant number of young people train in science and engineering and go into industry. Japan has been thin at the forefront of academic science but has established a system and a culture wherein a sizable percentage of young people gain scientific and technical training with an objective of going into industry.

The countries that have had economically successful leading industries have been strong across a wide spectrum of industries. One could read this statement as suggesting that strength in leading industries causes general economic strength. The inference I draw, however, is that the workings of a nation's basic economic institutions—those that determine its performance in education and in broad-gauged science, that support R&D and physical investment, and that achieve reallocations of labor and capital—have a broad atmospheric effect. If those workings are ineffective generally, it is unlikely that they will be effective for the high-technology industries. The contrast between Japan, Germany, and the United States—arguably the best economic performers in the postwar era—suggests that there is a wide range of viable institutional structures.

The most important lesson here is that nations aspiring to strength in high-technology industries had better attend to their general strength in technical education and establish and maintain a set of policies and institutions supporting general economic growth. A possible danger of the recent rhetoric about the importance of high-technology industries is that it may take attention away from these broader policy areas.

What Policies Seem to Have Worked?

There is no question, however, that industry-specific policies have had important effects. The preceding analysis revealed major differences in policies toward the three industry groups considered. Much of the current discussion of industrial policies seems to refer to the kinds of policies countries have directed toward their semiconductor and computer industries. Therefore, I will begin by focusing on these. Then I will turn to the lessons that might be drawn from the aircraft and nuclear power industries.

Lessons from Electronics-Oriented Policies. The United States and Japan clearly lead the pack in electronics, and both have had strong and effective policies supporting computers and semiconductors. The policies that resulted in American dominance in electronics after World War II were associated with national security programs. In Japan the policies that facilitated fast catching up have been associated with general MITI economic direction. Practically all analysts agree that these programs have had much to do with the two countries' success in electronics. Without trying to make these two obviously different policies appear the same, it nonetheless is worthwhile to search for common elements that perhaps can provide clues as to what kinds of policies are or are not effective. In fact, the policies have several elements in common.

Both programs involved a large protected home market. In the United Sates this was basically a government procurement market. In Japan, although the procurement market was far less consequential, the civilian market was also preserved for Japanese high-technology firms. Both the American military and the Japanese civilian markets were large enough so that several domestic firms could compete. In both cases the relevant government agencies were unwilling to set up a particular national champion. While the domestic industry has been sheltered from foreign competition, there has been vigorous internal competition, which has been the intent of those who have guided the policies.

This situation has had several ramifications. Maintenance of a domestic presence at the forefront of an industry was not dependent on the performance of any particular firm. In the industries in which backward or forward links were important, as in the computer and semiconductor industries, a firm was not locked into one supplier or one purchaser (except for the Department of Defense). And the strong demand for innovative products manifested in both markets motivated intense competition among domestic firms.

In both the United States and Japan publicly funded R&D programs significantly enhanced the abilities of the involved firms to produce advanced design products for commercial markets. In the Japanese case the principal programs involved support of generic research, done by company-employed scientists and engineers with the purpose of enhancing the company's technological strength relevant to commercial markets. In the United States the dominant programs were oriented to defense and to space exploration and involved both support of generic work and massive expenditures on hardware development. While not specifically intended to augment a company's commercial capabilities, this often was the result.

Put another way, although the two programs differed significantly in purpose and structure, each provided both a strong competitive market for domestic firms wherein technological prowess was rewarded and significant R&D support for firms in that market. In Japan stimulus of commercial competence was direct and intended, and in the United States commercial competence was created because military technology pulled civilian technology in its wake.

Much of the current discussion of policies in support of high-technology industries involves the term "picking winners." To what extent can the successful programs in the two countries be characterized in that way? If by that term one means sharply focused attention on achieving certain practical results, the proposition is apt. The U.S. programs of course were aimed at military objectives, not commercial ones; but the purpose certainly was to ensure a U.S. lead in the relevant technologies. Relatively clear-cut military hardware objectives lent a certain direction and thrust to the program of generic research as well as to that of hardware procurement. It should be recognized, however, that a central feature of the U.S. program was support of a wide range of options.

Picking and supporting winning industries in a commercial race might be an apt characterization of the Japanese programs, if the breadth of support is recognized. Thus semiconductors and certainly computers have been singled out for special attention; however, a wide range of electronics industries has received favored treatment.

Within particular industries and technologies, both the Department of Defense and MITI picked particular areas for intensive attention because of military potential in the former case and perceived potential commercial importance in the latter. In both cases particular companies or groups of companies were singled out for support. The big dollars in the U.S. program have gone to particular companies on R&D and procurement contracts. The DOD, of course, has not supported particular commercial ventures; and, contrary to some popular impressions, MITI has not in general tried to dictate to companies what kinds of products to design for sale on commercial markets. In both countries commercial competitive prowess has been enhanced through the strengthening of the design, development, and production capabilities of involved national firms, which in turn the firms used for what they judged to be commercially advantageous.

It is interesting to compare the U.S. and Japanese experiences with those of Britain and France. Although France, and to a lesser extent Britain, tried, neither of these countries established the same technology pull in their defense and space programs as did the United States. The total funds involved were vastly smaller. The efforts were less ambitious and were generally aimed at catching up with the Americans, not establishing new grounds. At the same time, the British and French programs have been prone to sink public funds into particular commercial designs. This approach has not been very fruitful in electronics. Although France has tried to protect its civilian market, its membership in the European market has forced it to be more open than Japan. In addition, branches of foreign-owned firms established within its own borders greatly complicated the business of even defining a domestic industry.

The generic research support programs of these countries have been much less coherently oriented than those of the United States and Japan. In France the commercially oriented aspects of the R&D support program tangled with the objective of establishing or preserving a French capability to design and produce military equipment. As a result clear commercial targets were not pursued, but the industry received shelter and subsidy simply to keep it operating. I already noted that the French military program's objective was only to stay close to the Americans, not to break radically new ground; hence little innovation has come out of it. Support of generic research in the British electronics industry, even aside from that associated with defense procurement, has not specially focused on areas judged commercially promising as has that support in the Japanese programs.

The effectiveness of the R&D support programs of the German BMFT, which both is more responsive to individual company proposals than the Japanese program and does support proprietary development work, has not, to my knowledge, been studied in any detail. It would be useful to know more about how this different kind of program is working out.

The new ESPRIT program, briefly described in chapter 5, may shove the Europeans off on a new path. The program in general, and the research in the particular areas being stressed, was developed in extended discussion with the major European firms in the information technology business. The focus of the program is on precommercial (generic) research, where the hope of gaining interfirm cooperation is not likely to founder immediately in questions of proprietary rights. The overall funding is small compared with total R&D in the European industry, but ESPRIT accounts for a significant fraction of funding of long-run generic research.

Lessons from Aircraft and Nuclear Power. Undoubtedly the great expenditures required to design and develop a particular new plane are the reason for such highly focused government attention and support. But, whatever the rationale, in the British and French aircraft industry the government has taken on the role of entrepreneur. For all the reasons cited earlier, this role is difficult for a government agency to play, and in most instances it has been played badly. Some learning, however, seems to have occurred. In the Airbus case the government(s) did a much better job of tapping commercially relevant expertise than has been done in earlier episodes. Instead of leading in its own preferred directions, in the Airbus case the governments organized, orchestrated, and subsidized a design and production cooperative closely tied to the articulated demands of the potential customers, the European airlines. The financial and organizational involvement of MITI in the development of the Boeing 767 also reveals keen attention to commercial promise.

Much more than government support of semiconductors and computers, government support of aircraft is readily identifiable with particular commercial products. The investments are far more lumpy. Aside from the giant American aircraft companies, private firms have shown reluctance to "bet the company" if they are not supported by their government. Thus the support programs are forced to aim for winners in a much narrower sense than support programs in electronics are. Programs supporting aircraft engines apparently also must aim for the most commercially successful firms. A consequence is that governments end up having a large financial stake in particular commercial products. Governments become partners with business, and partners having deep pockets.

As indicated, governments have displayed increasing sophistication about the importance of good market and technical analyses before placing bets. In the field of aviation, it is likely the lesson has been learned, and efforts such as the supersonic transport are behind us. But what is not clear is whether governments will learn when to cut losses. The Airbus may or may not yield a positive rate of return. The game certainly is chancy. But the involved governments do not act as if they could abandon the endeavor; they seem hooked politically to the programs. This inability to cut losses may be the most serious policy problem of support programs that involve huge lumpy public investments.

The nuclear power programs sharply reinforce these lessons. Like aircraft, nuclear power involves massive investments. While the nuclear programs are special in their intimate connection with noncommercial goals and values, they reveal vividly the problems that arise when a government commits itself to major investments in particular designs. The German and Japanese cases are noteworthy in that, from the beginning, the customers—the electric utilities—played a significant role in guiding R&D allocation. In the United States, Britain, and France, however, the lead government agencies made the decisions and simply presumed that the utilities would buy the reactors that were developed. It was a long time before those governments abandoned this policy.

As in the case of aircraft, clear evidence of learning exists in the area of nuclear power. In the United States the government cut the size of the reactor programs, and in Britain and France it reoriented the programs taking stronger account of economic calculus. As noted earlier, however, a recent study showed that only in Japan and France is reactor technology now more economic than power generators using fossil fuels. In all countries, the government has acted as if it had not only a huge stake in the reactor technologies it pushed but also endless funds and a reluctance to give up.

Summary. How does one summarize the lessons? What in them is germane to the present policy discussion?

The clearly powerful effects of the U.S. defense and space programs provide a complex and subtle message. These programs surely do not provide us with a model for future policies in support of high-technology industries. That U.S. procurement and procurement-related R&D had such a strong effect in building commercial leadership of U.S. firms certainly does not provide a persuasive argument that we should augment our present defense and space programs to increase "spillover." The massive expenditures we mounted then, and are incurring now, surely cannot be justified by the commercial returns.

It also seems likely that the large spillover from the defense and space programs of the late 1950s and 1960s was the product of a rather special set of circumstances. The military at that time greatly valued capabilities that could be realized through certain new technologies that were just emerging; and these capabilities, and the technologies more generally, also turned out to have great commercial value. Many analysts have suggested that spillover has diminished markedly since the mid-1960s.

The temptation will be to add commercial objectives to military ones in decisions about particular procurements or fields of R&D support. But one rather clear lesson of the post–World War II experience is that trying to blend commercial and military procurement objectives is a mistake. If a program is aimed specifically at enhancing competitive strength, it should stand separate from procurement-oriented programs.

MITI's programs are the best examples of relatively successful R&D support programs aimed specifically at creating a commercially competitive industry. In the following section I argue that a good case can be made for certain features of MITI-like programs, even in, or especially in, the United States. In thinking about such a transfer, however, I think it is wise to unpack the MITI experience.

The R&D support programs of MITI were complemented by considerable protection of the home industry and by a strong governmental role in picking the industries and designing the program. Protection is becoming increasingly difficult and fractious, even in Japan. In the United States proponents of active industrial policies offer them not as a complement for protection but as a substitute. The sharp industry targeting of the MITI programs was made possible because certain Japanese customs—which do not exist in the United States—were well established and because U.S. industry and technology provided a clear target for emulation. I suspect that, with no clear target established by other countries, MITI will now find deciding where to aim more difficult.

A policy of providing support for cooperative generic research, however, can be considered on its own merits. Such a policy seems well aimed at R&D in which the externalities are greatest; it seems welcomed rather than resisted by private industry; and it does not seem to involve government agencies in making judgments they are unequipped to make. Such a policy does not force a government agency to protect an industry or to make detailed commercial judgments.

Governmental involvement or partnership in the development, design, and production of particular commercial products poses a different set of issues, particularly if the costs are high and if there is room in the world economy for only a few competitive designs. I suspect that without this latter condition specific designs created under large government subsidy will not often play a major role in international competition. Private companies can afford to have a go at it on their own, and they have shown every inclination to be independent of government guidance or overview when they can. In particular, companies have a strong interest in keeping government away from their most promising product ideas. But when the costs of product development become very large in relation to the assets of even the biggest companies, it is a different game if a government stands ready to provide significant support.

And if that game is played out internationally with some products receiving major government subsidy, it is very fractious. If governments have learned enough to put SST's and gas-cooled reactors behind them and to place public money on designs that are reasonably attractive on international markets, they will soon have to learn what ground rules to place on heavily government-subsidized competition in high-technology products. This task may be difficult for diplomacy, and one can ask if the game is worth playing.

And this concern raises the next question.

Are Leading Industries Strategic?

The radical technological advances that we have seen in semiconductors, computers, aircraft, if not yet nuclear power, have had enormously wide ramifications. These surely are leading industries and technologies in the sense of Schumpeter. Although the fraction of national value added, employment, or capital stock contained in these industries has been quite small throughout the postwar era, these industries have shaped the new products that have emerged and the productivity growth that has been achieved in many other industries. Information processing, communications, and long-distance transportation of people have been revolutionized. And it is possible to trace the sources of this widespread economic revolution back to a very few leading industries.

But it is less obvious that leading industries have been strategic in the sense that the nations with strength in these industries have gained a widespread general advantage. Probably the United States was specially advantaged during the 1950s and through the mid-1960s. From the early 1950s, however, the other major industrial powers except Britain achieved much faster growth rates of productivity and of real per capita income than the United States achieved. To the extent that American R&D in the leading industries was an engine of basic growth, it certainly was pulling European and Japanese boats in our wake.

The European countries and Japan were especially concerned about technological gaps in the high-technology industries and clearly presumed that these gaps were strategically disadvantageous for them; however, it is not clear that their closing of the general productivity gap occurred because they closed the gap in high-technology industries. In all of these other countries, as in the United

States, the leading industries continue to account for only a relatively small fraction of value added and employment. That the fraction of exports attributable to leading industries is much smaller in Germany and Japan than in the United States, does not seem to have impeded German and Japanese industrial growth. For Japan exports of semiconductors are small compared with exports of automobiles. The value of German computer exports is swamped by the value of its chemicals and machinery exports. While Japanese excellence in the production of automobiles and motorcycles and German excellence in chemicals, machine tools, and related capital goods certainly rests considerably on technological sophistication in the high-technology industries, Japanese and German successes in these fields occurred before their semiconductor and computer industries began to challenge the American.

Increasingly, technological knowledge and capability are international rather than national. Except when governments block exports for national security or other reasons, competition in the high-technology industries is sufficiently strong that product trade rapidly makes available internationally products carrying the new technologies, regardless of where those products are made. The relevant scientific and technological communities are international, and generic knowledge spreads rapidly. The rise of first the American and then the European and the Japanese multinational corporation and more recently the surge of international joint ventures in R&D, design, and production of high-technology products have spread hands-on design and production capability among nations. The advanced industrial nations today are closely tied together technologically.

There are many reasons why this is so. One is the very nature of the leading industries of this half century: developments in communications and air transport have closed the distances between the technologically advanced nations. Another cause, ironically, is pursuit by governments of aggressive policies to ensure that their home industries keep up in high technology.

Because of strong competition, high-technology industries no longer seem necessarily to support especially high wages or rates of return on capital unless the industries are heavily subsidized. While technological advance and productivity growth in these industries are especially rapid, the gains go largely to those firms that buy the products of the industries, not to the firms in the industries. In the nature of the case, five countries cannot all be first in the product-cycle race, and the competition appears to have reduced the size of the prize and increased the costs of entry. The argument that leading industries are strategic nationally because they feed into national industries downstream is, to a considerable extent, vitiated by the growing strength and breadth of the international networks and the export orientation of the strongest firms in these industries. Indeed to push one's domestic industry and encourage home reliance upon it may disadvantage the closely linked industries rather than help them.

The questions here are difficult. My purpose is not to dismiss the proposition that leading industries are strategic for high-wage countries but to stimulate thought and research. Scholars of the semiconductor-computer interface attest that it calls for integrated companies or close intercompany relations. To a lesser degree the relation of airframe and engine designs is obviously close; a company in one area cannot proceed effectively without close interaction with a company in the other. The future will see more integrated companies and more tight intercompany relations. But the question I ask is whether or not national borders are strong hindrances to cross-country, intracompany (multinational) communication or ventures or to other close intercompany relations. I suspect the answer is "less and less," except insofar as national governments establish effective barriers.

A related issue is whether companies or countries can effectively pick and choose among aspects of a technology they seek to master or whether the connectedness of complex technological systems and developments means that one company or country must have competence across a broad front or be out of the game. Put another way, will dependence upon other companies or nations for critical parts of a system and specializing in certain aspects of a technology and ignoring the others diminish the ability of a company or a nation to keep up with new developments? Again, the answer may well be that, although companies do need broad-gauged competence to compete in complex, connected, fast-moving fields, companies in such fields are increasingly multinational. And joint ventures of a relatively broad and long-lived nature will become increasingly important, further blurring national lines.

Governments act as if they did not see clearly the increasingly international nature of capabilities in high-technology industries or as if, like King Canute, they could preserve their own technological islands from the seas. ESPRIT aims to create an expressly European technological capability. Philips and Inmos, as European companies, are involved with ESPRIT. But AT&T, now freed of the constraints of the old consent decree, is forging various collaborative ventures with Philips and is negotiating for an equity position in Inmos. Japan will try to build a Japanese aircraft industry but in collaboration with Boeing and Rolls Royce. The United States is now firmly in the game.

The U.S. Department of Defense is vigorously pursuing a policy of keeping American technology from Soviet hands. This policy has meant, on the one hand, placing various restrictions on exports to friendly countries in Europe and, on the other, keeping defense R&D contracts away from companies in the United States that have central offices abroad. But U.S. companies increasingly are engaged in technological interchange with European ones. And although Fairchild (now owned by a French-based company) can be excluded from the VHSIC program, Fairchild will continue to hire scientists and engineers away from companies that are involved. The traffic between islands cannot be stopped or even much curtailed. King Canute could not keep the waves from rolling high on the shore.

7
What Implications for U.S. Policy?

Devising new policies that are effective and not fractious is likely to be difficult and frustrating. And the stakes may not be as high as is often argued. Nonetheless, some rethinking or some fresh thinking clearly is in order. Such thinking should not presume, however, that we can start with a clean slate. U.S. policies supporting high-technology industries will almost surely continue to be heavily influenced by national security interests. More than any other of the major industrial nations, the U.S. government will continue to be constrained in the modes of interaction with industry that will be politically acceptable.

The National Security Connection

Commentators such as Magaziner and Reich (1982) have argued that U.S. policies toward high-technology industries, traditionally based as they are in defense procurement interests, provide less economic advantage than would more commercially oriented policies, and they have proposed that we establish policies more explicitly aimed at economic objectives. Although such a proposal might be a good idea, to forget about the national security connection would be a mistake. In the past, defense-oriented policies have had enormous commercial effect, and in many areas there continues to be significant spillover. Even under the most optimistic assumptions about arms control, military procurement and procurement-related R&D will, for the foreseeable future, continue to be by far the major source of government support for high-technology industries in the United States. And, in these industries, national security considerations will strongly resist being cleanly separated from economic ones.

The high-technology industries are inextricably connected with perceptions of national security and vulnerability. As we have seen, in nations with a significant military procurement program it is hard to draw clean lines between procurement policies and industrial policies. At least until recently, the U.S. government did not explicitly concern itself with the commercial strength of the firms in its defense industries, but perhaps it did not because those firms were doing well in commercial markets so there was nothing to worry about. Because of our status as the arsenal of democracy, the United States will continue to spend significant funds on R&D in these industries, enough so that we do not perceive ourselves as lagging technologically in any important area.

What is new in the current context is that the technological threats we see may come more from our allies than from the Soviet Union and may appear in the form of commercial products rather than that of weaponry. I suspect that, try as we may to distinguish between preeminent military design capability and commercial success in high-technology industries, these aspects will blur in people's minds. If the Japanese can build a fifth-generation computer before an American firm can, confidence that we are at the top of the field for military application surely will be undermined. Both symbolic and real elements are involved; statements emanating from the Department of Defense have clearly admitted as much.

For all these reasons, our policies in support of high-technology industries will continue to be intertwined with national security objectives.[17] Indeed, such intertwining may be a political requirement for significant government support.

We should understand, however, that other countries will accuse us of having major industrial policies disguised as military R&D programs. Europeans, pressed on the fairness of their express industrial support programs, long have responded that we have done much more under DOD auspices. As Department of Defense R&D support programs become identified with matching or beating the explicitly commercial programs of other countries, the flack will get thicker. We will lose much of whatever credibility we now have in arguing that the programs of other countries amount to unfair subsidization. Also, to the extent the Department of Defense continues to press for keeping American technology out of foreign hands, we lose force in arguing for other countries to open up their R&D support programs to American companies.

This issue is a fractious one, setting at odds not only the United States and friendly nations but also departments within the U.S. government. The practice of vigorous and widespread control of high-technology exports, to friendly countries as well as to the Soviet Union, is troublesome. Perhaps even more troublesome are the attempts to require clearance for reports of academic research supported by government, even when the research was not preclassified as secret. These policies anger people and nations afflicted by them, and they are going to frustrate those people and nations that invoke them. King Canute knew he could not hold off the waves, and his was only a gesture. But the U.S. government, or parts of it, seems to be extremely serious in its attempt.

In any case, military and civilian technology inevitably will be tangled. As a result, it is a good bet that R&D and procurement related to national security will suffice to keep American firms at the technological forefront in commercial as well as military technology, if these are at all connected. When commercial demands have little contact with plausible military needs, however, pressure will build for new policies. It makes sense, therefore, to begin to think of a set of complementary policies more explicitly oriented toward economic objectives. As we ponder such policies, both the particular American political context, which surely constrains our actions, and the issue of international policy conflict ought to be considered carefully.

Support of Generic Research Done by Industry

In my opinion, support of generic research done by industry is the most promising way to go. This type of R&D support by MITI appears to have been most effective. Until now the Department of Defense and NASA have been virtually the only governmental supporters of such work in the United States. There are strong reasons for establishing a basis of support independent of the Department of Defense.

American companies now are strongly indicating that they would like to band together to fund cooperative generic research, even in industries in which the Department of Defense substantially finances such work and even when no public funds are provided to catalyze the effort. In particular, several of our semiconductor and computer manufacturers have already joined together to do such research through the newly formed Microelectronics and Computer Technology Corporation.

The Department of Justice, in a preliminary ruling, has indicated that it does not see any antitrust issues at stake so long as the supported research stays generic in nature. The proprietary interests of the involved companies probably will ensure that this cooperative endeavor not venture too close to the individual companies' potential proprietary interests.

The American business community has reason to be nervous, however, and there is good reason to get explicit changes in the antitrust laws into place. The law is ambiguous. Advisory rulings by the antitrust division are not binding, not even on itself, particularly if leadership changes; and in any case such rulings cannot block private litigation. The triple damages penalty, plus the ability of other affected parties to piggyback on a successful suit by one, can loom as a real threat. A possible model for building in more protection against private suits is provided by the recently passed Export Trading Company Act. Under this model, private companies would be required to disclose their proposed joint venture activities. If no significant antitrust issues were found in the proposal, the venture would be considered cleared. Private suits would not be barred; but anyone winning such a suit would collect only single damages, and those losing private suits would be assessed court costs. Such a legislative act should clear the air regarding cooperative generic R&D ventures.

Given American traditions, it would appear that industry, with government encouragement, should lead in initiating such programs, and government should make no attempt to force the program or to direct it in any detail. The Cooperative Automotive Research Program, initiated under the Carter administration and aborted under the Reagan administration, was for support of generic research of the kind discussed here. The automobile companies, however, had no part in initiating or designing the program and felt it was being forced on them. The program might have gone quite differently had the automobile companies been urged to design it for themselves.

As the Microelectronics and Computer Technology Corporation indicates, the private firms may be willing to invest considerable amounts of their own money in such cooperative generic research programs. I would endorse the idea of having government funds supplement those funds, however. Such public financial assistance might be provided on a formula basis, as through the provision of matching funds. Alternatively, the decision about whether or not to provide public support might be made on a case by case basis, although I am uncomfortable with the political and organizational problems that such a policy would engender. Government identification of industries that can benefit from cooperative generic R&D seems a problematic venture, particularly in the United States. There is the real danger of special interest pork barrel politics. Also at the present time we do not have an executive department whose judgments in these matters seem worthy of trust and resistant to manipulation. Thus it seems better to let the initiative come from industry.

One important policy issue regarding such generic research cooperatives that is sure to arise involves the terms of exclusion from such groups. This issue is delicate. A generic research cooperative that involves, say, the three largest firms in an industry and excludes others ought to be ruled in violation of the antitrust laws. In such endeavors that involve no public funding, however, I would argue a rule of reason.

For cooperative generic research groups employing public funding, I take a different stand. I believe it is in the interest of the United States and of all countries together that participation in publicly subsidized programs be open to all companies with an R&D and a production presence in the sponsoring nation. I would propose that U.S. government-funded programs of this sort be open to foreign firms, *provided* reciprocity is shown by a firm's home government on comparable programs. This, of course, is another argument for sponsoring these programs in the United States through a vehicle other than the Department of Defense. Getting other countries to abide by these ground rules will not always be easy, but the pursuit of reciprocity provides one useful guide star for American diplomacy. A significant program of government-funded cooperative generic research, backed by a reciprocity policy, has promise of giving us leverage on the programs of other countries, most notably those of Japan, that we presently do not have.

Direct and Indirect Support of Commercial Design and Development

The issues of reserved or protected markets and of government subsidization of particular commercial products always have been extremely conflictive aspects of national policies supporting high-technology industries. And these aspects promise to cause considerable international conflict and economic waste in the future unless they are somehow controlled.

In fact, the increasing internationalization of technology, and in particular the growing proclivity of companies to undertake joint ventures, already is visibly undermining these traditional national policies. Also, the most important of the previously closed civilian markets, that of Japan, is slowly and painfully opening up, at least to a degree. The rising international competition in high-technology products, however, is sure to threaten weak national industries and invoke import barriers of various sorts. One can recall the recent French blockage of Japanese video cassette recorders. But also recall that the result of that conflict was a joint venture with a Japanese firm. The U.S. high-technology industries have not been shy about requesting protection, as they showed when Japanese televisions flooded the U.S. market. Such a response could occur with semiconductors, computers, or aircraft.

I think it important that the United States take a strong position against protectionist policies but not be sanctimonious about it. The United States probably will be on better grounds arguing against general protection than against procurement policies that cater to national firms. Among other things, we undoubtedly will preserve the largest protected procurement market in the world—that tied to our defense budget. We should not be surprised if our arguments to other countries that they should open up telecommunications equipment procurement are met with the reminder that what is sauce for the goose is sauce for the gander. It is likely that the firms themselves, except possibly in Japan, increasingly will frustrate procurement policies oriented toward home firms by joining together in ventures.

The growing tendency of firms in different countries to band together in joint ventures on large, expensive projects is also likely to complicate national efforts to help home industry by R&D or general subsidy. But such efforts, encouraged and subsidized by governments, are likely to become increasingly common. I see no reason to believe that government agencies will greatly improve their ability to pick winners. If the lessons of Airbus are heeded, however, there may be less of a proclivity to support big losers that cannot seriously compete even with heavy continuing subsidy. While the rate of return to the European countries on Airbus is likely to be low if not negative, that plane is competitive on world markets, at least with the subsidies governments seem willing to provide. The American companies clearly feel they were unfairly hurt by Airbus. What should be our policy in similar cases in the future?

I think it is important to distinguish three different aspects of the Airbus program. First, under governmental auspices a consortium of companies was organized to work together on a major commercial product. Second, significant government funding certainly was involved in the research and development stage and probably in production as well. Third, the governments of the major countries that participated in the Airbus consortium attempted to pressure their national airlines to buy Airbus. In my view the second and third aspects of the Airbus experience should be sharply distinguished from the first.

At present U.S. policy seems ambivalent regarding design and production joint ventures. The U.S. antitrust laws currently are not being interpreted as ruling out cooperation among American companies that produce different components of a system—as between an airframe manufacturer and engine producer. Nor have they been used to rule out joint ventures between an American firm and a foreign firm in the same line of business. As I understand the matter, there is no explicit ruling that international joint ventures of American companies do not infringe on antitrust law, only the fact that to date no such venture has been successfully attacked, and there is no established case law. This situation may be a time bomb.

It seems time to rethink the whole issue. Although I am less easy about joint design and production ventures than I am about generic research cooperation, it does not seem right that such an international venture would receive totally different treatment from that received by two or more U.S. firms in a design and production venture for which the market is clearly international. Actually, the issue is delicate. On one hand, international consortia have special advantages, in that they make it difficult for governments that subsidize or provide protected markets to aim their policies toward home firms only. On the other hand, there surely is an issue here akin to the older ones about trade creation and trade diversion. The United States, perhaps with Japan, is in a special position in that in industries in which such consortia may be common, we often have several firms. Thus our firms have the opportunity to look for national not just foreign partners. It seems odd that we would discriminate against a national partnership if each partner judged this venture more promising economically than an international consortium.

There is an even more basic question: What stance should the United States take when there is an obvious trade-off between the number of rivals and the degree of wasteful overlap of effort? The United States or the world is not necessarily better off having two or three similar but competitive U.S. designs when several foreign designs are available and significant fixed costs are involved in each entry in the market. Although I would distrust a U.S. governmental authority that initiated design and production partnerships of this sort, I suggest that it is time to loosen the present strictures against design and product cooperation by U.S. firms under certain circumstances. The circumstances are, first, the existence of competitive foreign efforts and second, the presence of such significant fixed costs per entry that the market presumably can support only a few designs. It seems prudent to pass laws, similar to those in the Export Trading Company Act, regarding such ventures.

Again, I am suspicious about machinery to let government initiate such ventures. I would rather have a public stance of careful scrutinization of private proposals.

In the matter of subsidization by foreign governments of particular designs, I think the U.S. government should take a strong stand that, so far as the U.S. market is concerned, such subsidy is unfair competition. As is our right under GATT, we should stand ready to impose tariffs commensurate with the degree of subsidization. While calculating the degree of subsidy is a complicated business with perhaps no right answer, I think that the United States should advertise its intention to offset the advantages of foreign subsidies when the competition is in our home markets.

I think it prudent to separate the issue of whether we should try to offset the effect of foreign subsidization in competition for the U.S. domestic market from that of how we treat subsidized or protected foreign markets. The former is under our direct control; the latter is not. Although we can bend our negotiation efforts to opening up foreign markets and reducing the degree of discrimination in favor of home companies, that can be a hard row to plow.

Reprise

The guidelines for new policies sketched above certainly will seem insipid to those who are looking for bold new departures. They certainly seem weak tea compared to those that other countries have put in place or those that have been discussed by advocates of a far more activist industrial policy for the United States.

But if the description and analysis presented in this study are close to the mark, there is not much about the active industrial policies of other countries that we ought to be emulating. For the most part, the foreign record has been one of expensive frustration. Other countries keep trying active policies in support of their high-technology industries, not because past policies have been deemed successful but because the high-technology industries continue to be weak and there are strong national urges to do something about that weakness.

The exception, of course, is Japan. I have stressed, however, that many attributes of Japan have contributed to its remarkable (until recently) economic performance and that assessing the importance of its industrial policies is hard. In any case, MITI must be understood as part of a package of political institutions and cultural predispositions. Although earlier I argued that MITI likely will in the future have greater difficulty targeting industries than when the United States provided a clear model, I believe MITI will continue to play an active role in Japan. Policies strongly favoring certain classes of industry and providing considerable if broad-gauged industry guidance, however, simply will not be accepted in the United States unless they are tied, in a real or a symbolic way, to national security. And, if the national security connection is largely symbolic, the likely result will be either a new project Apollo or a pork barrel, but almost certainly not a policy that looks like MITI's. This situation may be a liability or an advantage, but for the foreseeable future it is a fact.[18]

We need to pick and choose from the policies that have been piloted by other countries, considering seriously only those that have showed promise abroad and look as if they might be implemented effectively here. I have given my judgments of what those policies are.

We need to pay more attention to our assets in the race. U.S. defense R&D expenditures will continue to dwarf those of our industrial competitors. While in some areas military R&D may have little to do with the creation of commercially relevant technologies, military R&D and procurement will, for better or for worse, be the dominant specific influence on our high-technology industries. I believe that this influence will continue to keep American firms competitive commercially in those areas that are close to military interests.

The United States has had, at least until recently, the broadest gauged educational system in the world, and we still have a significantly higher percentage of young people going on to postsecondary education than anywhere else except Japan. The economy of the United States has an internal competitiveness and openness to new ideas and new firms that none of our industrial competitors presently is close to matching. Our policies should exploit these advantages and not let them erode.

There certainly is reason to focus attention to the broad-gauged educational front, an area that may be overlooked if we start with the premise that industry-specific policies are the key to success. It is hard to say if expressed concerns about inadequate supplies of young well-trained engineers and applied scientists in central fields are overblown; but apparently we have worked ourselves into a position in which the university departments training the needed people are short of faculty, in part because nonacademic jobs are more lucrative. Perhaps the time is again ripe for the large public programs in support of higher scientific and technical education that marked the post-Sputnik period. The indications are, however, that our educational problem is much deeper than appears when one looks only at advanced training. Over the long run, improving the teaching of science and mathematics in primary and secondary schools may be more important in preserving an American lead in the high-technology industries of the future than creating specific programs aimed at a narrow front of our high-technology industries today.

And it is exactly the internal competitiveness of U.S. industry that makes policies that are appropriate or even needed in other countries infeasible and counterproductive here. Throughout this essay I have stressed that the Schumpeterian engine of progress involves public as well as private components. On one hand, not to see the importance of public institutions is intellectual nearsightedness on the part of many advocates of free enterprise. On the other hand, a weakness of many recent arguments in favor of industrial policy is the failure to understand how Schumpeterian competition works and what its strengths and its limitations are. Industrial policy in the United States needs to be well designed to alleviate the limitations, without hindering the strengths.

Many years ago, in his *Capitalism, Socialism, and Democracy* (1942), Schumpeter took the position that modern man was close to routinizing the innovation process. He felt that rational calculation and discussion were eliminating from it uncertainties and divergencies of judgment and that the hurly-burly of capitalist competition, which he acceded had been a fount of creativity and energy, if lost would not be missed. This forecast seems false. The United States may be handicapped in relation to other countries in the extent to which efforts at innovation can be coordinated. This lack of coordination may hurt us in some areas, particularly those in which the costs of the endeavors drive out much chance for sustaining several approaches. The sheer size of our corporations and our internal market, however, may help us avoid being closed down in these areas if we adopt sensible policies. And in most areas economies of scale are not so overwhelming. The U.S. economy continues to have an openness to entry of new firms and new ideas that other countries do not and that they increasingly seem to discourage in the name of industrial policy.

If MITI does not seem likely in our future, the flexible industrial structure of the United States should not be discounted as a formidable competitive engine of progress. We may be lucky that it so stubbornly resists being targeted, coordinated, or planned.

Notes

1. The most sophisticated of the recent statements are by Baranson and Malmgren (1981), Magaziner and Reich (1982), and Zysman and Tyson (1983).

2. Thomas and Jennings Piekarz (1983) present an analysis of the R&D statistics similar to mine. They also hazard some comparisons across countries in such variables as tax treatment of R&D spending.

3. For a heroic attempt to assess the role of "advances in knowledge" (not explicitly R&D) in the productivity growth experience of different countries see the work of Edward Denison (1967 and 1976).

4. Stein and Lee (1977) have provided the best study I know about differential productivity growth rates across countries at the sectoral level.

5. One of the best early studies was that by Nestor Terleckyj (1974). Edwin Mansfield's more recent study (1980) divides R&D into basic and applied and into privately and publicly financed.

6. Terleckyj (1974) and Mansfield (1980) are representative of studies that treat private and public R&D as (logarithmically) separate factors of production. Link (1981) and Kalos (1983) treat public R&D as affecting the productivity of private R&D. Kalos provides a good review of this literature.

7. Robert Evenson (1982) has recently ably reviewed that literature.

8. See *Mathematica Inc.* (1976).

9. The following draws from a number of sources, and partly recapitulates my earlier discussion of U.S. policy in Nelson, Peck, and Kalachek (1967) and Nelson (1982).

10. The following draws from various sources. See in particular Vernon (1974), Gershenkron (1962), Pavitt (1980), Pavitt (1976), Rothwell and Zegveld (1981), Warnecke and Suleiman (1975), and Katzenstein (1978).

11. My principal references for this account are Patrick and Rosovsky (1976) and Johnson (1982).

12. This account of U.S. policy toward semiconductors and computers draws heavily on the essays by Levin and by Katz and Phillips in Nelson (1982). See also Wilson, Ashton, and Egan (1980) and Kalos (1983).

13. The following discussion of the European experience draws in particular from Sciberris in Pavitt (1980), Zysman (1977), Dosi (1981), and Malerba (1983). I am especially indebted to Franco Malerba for helping me understand the European record.

14. The following discussion is based principally on the following sources: Peck and Wilson (1982), Peck (1983), Pugel, Kimara, and Hawkins (1983), Wheeler et al., (1983), and Doane (1983). I am particularly indebted to Donna Doane for having made available to me her draft manuscript.

15. My principal source for the U.S. study was the chapter by Mowery and Rosenberg in Nelson (1982). William Spitz collected the materials on the European experience in this paper ''European Policies in Support of the Civil Aviation Industry.'' The Airbus story was drawn in part from Newhouse (1982) and from Lorell (1980). The analysis of the Japanese case is drawn in good part from Mowery and Rosenberg (1984).

16. The most important sources of the following discussion were Walker and Lönnroth (1983), Keck (1981), Suttmeier (1982) and Hazelrigg and Roth (1983). Michael Sullivan ably surveyed the European experience.

17. The ''super computer'' project of the Department of Defense is a good example of a program triggered in part by the perception that a friendly country (Japan) might get ahead of the United States in a technology perhaps relevant to national security.

18. For a compatible view see Schuck (1983).

Bibliography

Adams, J. G., and L. R. Klein. *Industrial Policies for Growth and Competitiveness.* Lexington: Lexington Books, 1982.

Baranson, J., and H. Malmgren. *Technology and Trade Policy.* Washington, D.C.: Malmgren, Inc., 1981.

Denison, E., with W. Chang. *How Japan's Economy Grew so Fast.* Washington, D.C.: The Brookings Institution, 1976.

Denison, E., with J. Poullier. *Why Growth Rates Differ.* Washington, D.C.: The Brookings Institution, 1967.

Doane, D. L. ''The Generation of New Products and Industries.'' Draft manuscript, Yale University, 1983.

Dosi, G. *Technical Change and Survival: Europe's Semi-Conductor Industry.* Sussex European Research Center, University of Sussex, 1981.

Eads, G., and R. R. Nelson. ''Government Support of Advanced Civilian Technology: Power Reactors and the Supersonic Transport.'' *Public Policy,* Summer 1971.

Evenson, R. ''Technical Change in U.S. Agriculture.'' In R. Nelson (ed.), *Public Policy and Technical Progress: A Cross Industry Analysis.* New York: Pergamon Press, 1982.

Flaherty, T., and H. Itami. ''Financial Institutions and Financing for the Semiconductor Race.'' In D. Okimoto (ed.), *Competitive Edge.* Stanford University Press, (forthcoming).

Friebe, K. P., ''Industrial Policy in the Federal Republic of Germany.'' In K. P. Friebe and A. Gerybadze (eds.), *Microelectronics in Western Europe.* Berlin: Erich Schmidt Verlag, 1984.

Gershenkron, A. *Economic Development in Historical Perspective.* Cambridge: Harvard University Press, 1962.

Griliches, Z. ''Returns to Research and Development Expenditures in the Private Sector.'' In J. Kendrick and B. Vaccara (eds.), *New Developments in Productivity Management.* New York: NBER, 1977.

Hazelrigg, G., and E. Roth. *Windows for Innovation: A Story of Two Large Scale Technologies.* Submitted to the NSF by Econ Inc., Grant #PRA-8110724, April 30, 1983.

Johnson, C. *MITI and the Japanese Miracle: The Growth of Industrial*

Policy 1925–1975. Stanford: Stanford University Press, 1982.

Kalos, S. *The Economic Impacts of Government Research and Procurement: The Semi-Conductor Experience.* Yale University Ph.D. Dissertation, 1983.

Katzenstein, P. (ed.) *Between Power and Plenty: Foreign Economic Policies of Advanced Industrial States.* Madison: University of Wisconsin Press, 1978.

Keck, O. *Policymaking in a Nuclear Program.* Lexington: Lexington Books, 1981.

Landis, D. *The Unbound Prometheus.* Cambridge: Cambridge University Press, 1970.

Link, A. *Research and Development Activity in U.S. Manufacturing.* New York: Praeger, 1981.

Lorell, M. A. *Multinational Development of Large Aircraft: The European Experience.* The Rand Corporation (R-2596 DR and E), Santa Monica, July 1980.

Magaziner, I., and R. Reich. *Minding America's Business: The Decline and Rise of the American Economy.* New York: Vintage Books of Random House, 1982.

Malerba, F. *Technical Change, Market Structure, and Government Policy: The Evolution of the European Semi-Conductor Industry.* Yale University Ph.D. Dissertation, 1983.

Mansfield, E. "Basic Research and Productivity Increase in Manufacturing." *American Economic Review,* December 1980.

Mathematica Inc. *Quantifying the Benefits to the National Economy from Secondary Applications of NASA Technology.* NASA Contractor Report (R-2674), Washington, D.C., 1976.

Mowery, D., and N. Rosenberg. "The Commercial Aircraft Industry." In R. Nelson (ed.), *Government Support of Technical Progress: A Cross Industry Analysis.* New York: Pergamon Press, 1982.

Mowery, D., and N. Rosenberg. "Government Policy, Technical Change, and Industrial Structure: The U.S. and Japanese Commercial Aircraft Industries, 1945–83." Mimeo prepared for the U.S.-Japan High Technology Research Project Conference, Stanford University, March 21–23, 1984.

Nelson, R. (ed.). *Government Support of Technical Progress: A Cross Industry Analysis.* New York: Pergamon Press, 1982.

Nelson, R., M. J. Peck, and E. Kalachek. *Technology, Economic Growth, and Public Policy.* Washington: The Brookings Institution, 1967.

Newhouse, J. *The Sporty Game.* New York: Alfred Knopf, 1982.

Patrick, H., and H. Rosovsky (eds.). *Asia's New Giant: How the Japanese Economy Works.* Washington: The Brookings Institution, 1976.

Pavitt, K. ''Government Support for Industrial Research and Development in France: Theory and Practice.'' *Minerva,* Autumn 1976.

Pavitt, K., (ed.) *Technical Innovation and British Economic Performance.* London: Macmillan, 1980.

Peck, M. J., ''Government Coordination of R&D in the Japanese Electronics Industry.'' Yale University (mimeo), 1983.

Peck, M. J., and R. Wilson. ''Innovation, Imitation, and Comparative Advantage: The Performance of Japanese Color Television Set Producers in the U.S. Market.'' In H. Giersch (ed.), *Emerging Technologies: Consequences for Economic Growth, Structural Change, and Employment.* J. C. B. Mohr, Tubingen, 1982.

Piekarz, R., E. Thomas, and D. Jennings. ''International Comparisons of Research and Development Expenditures.'' Division of Policy Research and Analysis, National Science Foundation (mimeo), January 6, 1983.

Pollard, S. *Peaceful Conquest: The Industrialization of Europe 1760–1970.* Oxford: Oxford University Press, 1981.

Pugel, T. A., Y. Kimara, and R. G. Hawkins. ''Semi-Conductors and Computers: Emerging International Competitive Battlegrounds.'' In R. W. Moxen, R. W. Roehl, and J. S. Truitt (eds.), *International Business Strategies in the Asia-Pacific Region.* JAI Press, 1983.

Rothwell, R., and W. Zegveld. *Industrial Innovation and Public Policy.* London: Frances Pinter, 1981.

Schuck, P. ''Industrial Policy's Obstacles.'' *The New York Times,* September 7, 1983.

Schumpeter, J., *Capitalism, Socialism, and Democracy.* New York: Harper and Brothers, 1942.

Schumpeter, J. *Business Cycles.* New York: McGraw Hill, 1939.

Stein, J., and A. Lee. *Productivity Growth in Industrial Countries at the Sectoral Level 1963–1974.* Santa Monica: The Rand Corporation (R-2203-CIEP), 1977.

Suttmeier, R. P. ''The Japanese Nuclear Power Option: Technological Promise and Social Limitations.'' In R.A. Morse (ed.), *The Politics of Japan's Energy Strategy.* Institute of East Asia Studies of the University of California, Berkeley, 1982.

Terleckyj, N. *The Effects of R&D on the Productivity Growth of Industries.* Washington, D.C.: National Planning Association, 1974.

Vernon, R., (ed.). *Big Business and the State.* Harvard University Press,

1974.
Walker, W., and M. Lönnroth. *Nuclear Power Struggles: Industrial Competition and Proliferation Control.* London: George Allen and Unwin, 1983.
Wallerstein, M. B. ''Scientific Communications and National Security in 1984.'' *Science,* May 4, 1984.
Warnecke, S., and S. Suleiman. *Industrial Policies for Western Europe.* New York: Praeger, 1975.
Wheeler, J., M. Janow, and T. Pepper, *Japan's Industrial Development Policies in the 1980's.* Hudson Institute, Croton-on-Hudson, New York, 1982.
Wilson, R. W., P. K. Ashton, and T. P. Egan. *Innovation, Competition, and Government Policy in the Semi-Conductor Industry.* Lexington: Lexington Books, 1980.
Zysman, J. *Political Strategies for Industrial Order: State Power and Industry in France.* Berkeley: University of California Press, 1977.
Zysman, J. *Governments, Markets, and Growth: Financial Systems and the Politics of Industrial Change.* Cornell University Press, 1983.
Zysman, J., and L. Tyson. *American Industry in International Competition: Government Policies and Corporate Strategies.* Ithaca: Cornell University Press, 1983.

20